Membrane Reactor Technology

Rakesh Govind and
Naotsugu Itoh, editor

Gerardo Catapano
A. Champagnie
Philippe Cini
G.R. Gavalas
Rakesh Govind
Michael P. Harold
H.P. Hsieh
Shamsuddin Ilias
Naotsugu Itoh
Elias Klein
S.W. Nam
N. Nourrbaksh
Bruce Patenaude
T.T. Tsotsis
Venkataraman
Richard A. Ward
Terry E. Williams

AIChE Symposium Series
Number 268 1989 Volume 85
Published by
American Institute of Chemical Engineers
345 East 47 Street New York, New York 10017

AIChE shall not be responsible for statements or opinions advanced in papers or printed in its publications.

Library of Congress Cataloging-in-Publication Data
Membrane reactor technology.

(AIChE symposium series ; no. 268, v. 85)
Selected papers presented at the AIChE Annual Meeting in Washington, D.C., Nov. 1988.

Includes index.

1. Membrane reactors—Congresses. I. Rakesh Govind. . II. Itoh, Naotsugu. III. American Institute of Chemical Engineers. Meeting (1988 : Washington, D.C.) IV. Series: AIChE symposium series ; no. 268.
TP248.25.M45M45 1989 660′.283 89-14908
ISBN 0-8169-0464-2

FOREWORD

The term "membrane reactor" has been used in the literature to describe a number of reactor configurations. In some instances, the role of the membrane is to simply provide a barrier to confine the catalyst to the reactor or prevent certain components, such as particulates, from contacting the catalyst. In some applications, the membrane is used as a support material for the catalyst. Membrane reactors can be classified as follows:

Type of reaction

homogeneous or catalytic

Integration of membrane and catalyst functions

membrane separated from reactor
— purification of reactant and/or product

membrane as part of reactor but not catalyst
— catalyst packed on one side of membrane
— catalyst deposited as a thin film on the surface

membrane as prart of catalyst/catalyst support
— catalyst deposited inside the membrane pores
— membrane material is catalytic
— catalyst (liquid) impregnated inside the membrane pores

Membrane configuration

plate and frame, spiral wound, tubular or hollow fiber

Component permeating or flowing through membrane

reactant permeating through membrane as in membrane catalysis

reactant flowing through membrane for uniform distribution of reactant

product permeating through membrane to shift the reaction from equilibrium or reduce product inhibition

Flow patterns on either side of membrane

co-current (flow on both sides of the membrane in the same direction)
counter-current flow

Structure of membrante

large pores, not permselective

microporous, permselective (molecular sieving, capillary condensation with liquid flow, surface diffusion, knudsen diffusion)

dense, permselective (diffusion and solubility)

Membrane reactors offer several advantages, such as short contact time, controlled pore size and distribution, high surface/volume ratio, less mass transfer resistance, and simultaneous reaction with separation. However, the development of membrane reactors, at least for catalytic reactions, requires the availability of membranes which can withstand high temperatures. Incorporating permeselectivity for gas separation into such high temperature membranes is even more challenging. Current efforts in developing such membranes have utilized expertise in ceramic science, aerosol reactors, chemical vapor deposition, development of thin films, material science and chemical engineering.

In this symposium series, a balance of papers have been included on the development of membranes and in the various applications of membrane reactor technology. The field of membrane reactor technology is in its infancy and significant effort will be needed before it can find industrial application.

Rakesh Govind, *editor*
Naotsugu Itoh

CONTENTS

REDUCTION OF PRODUCT INHIBITION BY USE OF AN ENZYME MEMBRANE REACTOR IN THE PROCESSING OF NATURAL SUBSTRATES

Terry E. Williams, Gerardo Catapano*, Elias Klein and Richard A. Ward ■ Division of Nephrology, School of Medicine, University of Louisville, Louisville, KY 40292

An enzyme reactor coupled to a dialyzer and a reverse osmosis unit was used to continuously produce concentrated low molecular weight products. An integral model was developed to describe mass transfer phenomena in the reactor system. The performance of the reactor was evaluated using enzymes operating on natural substrates. The reactor system allowed control of the molecular weight distribution of the products. In the case of reactions which are strongly inhibited by low molecular weight products, reaction rates were found to be three orders of magnitude higher than in a batch system.

INTRODUCTION

Enzyme Membrane Reactors (EMRs) provide an alternative to traditional processes in the production of specialty chemicals. Because enzyme molecular weights are usually in the range 10000 to 100000, ultrafilters (UFs) have generally been used in EMRs. However, UFs allow only poor separation of low molecular weight products (<1000) from reaction intermediates in the 10000 to 40000 molecular weight range. In contrast to ultrafiltration (UF), dialysis separates on the basis of a compound's diffusivity through a membrane (1). The strong dependence of the diffusion coefficient on molecular weight (2) allows dialysis to separate compounds with molecular weights below about 2000 from higher molecular weight species. The disadvantage of dialysis is that a highly diluted product is obtained. However, by coupling a reverse osmosis (RO) unit with a dialyzer in a closed loop, the products can be concentrated and dialysate simultaneously regenerated for further use.

In this paper, we describe the performance of an EMR using a dialysis module connected in closed loop with a RO unit to transform natural substrates with hydrolytic enzymes. Two different enzyme systems were used.

a) A mixture of endopeptidases, trypsin and chymotrypsin, was used to convert cheese whey into peptides of about 1000 molecular weight. These peptides are intended as a replacement for glucose in peritoneal dialysate formulations (3).

b) An exoglucosidase, β-amylase, was used to catalyze maltose production from soluble starch.

The transient kinetic behavior of the EMR was modelled in terms of integral mass balances about each reactor component, and the model was tested by tracing experiments.

MATERIALS AND METHODS

The reactor system is shown in Figure 1. Substrate is contained in the reactor, which is an isothermal stirred-tank. Hydrolysis is started by addition of enzyme to the reactor. The contents of the reactor are pumped to a hollow-fiber dialyzer. Unreacted substrate and enzyme are retained in the dialyzer and return to the reactor where further hydrolysis may occur. Product species small enough to permeate the wall of the fibers diffuse down their concentration gradients towards the shell side of the dialyzer and into the well mixed accumulator. From there, fluid containing product species is pumped to the RO unit. Since the RO membrane is essentially impermeable to dissolved solutes, only water traverses the RO membrane to form dialysate for the dialyzer. Small molecular weight products return to the accumulator.

*Author to whom correspondence is to be addressed.

For experiments with the endopeptidases, two dialyzers with cellulose membranes were used. The first, CF1211L (Baxter Healthcare, Chicago, IL), had low permeability membranes, while the second, D-6 (Fresenius, Bad Homburg, FRG), had high permeability membranes. A dialyzer with high permeability cellulose membranes, BLX 16 GW (Nikkiso, Tokyo, Japan), was used for the experiments with β-amylase. The mass-transfer properties of these hollow-fiber dialyzers were determined in separate experiments using a 5%, by weight, mixture of glucose oligomers (4). The same method was used to determine the mass-transfer coefficients of peptide species produced by the reactor system, with a 5% (by weight) solution of freeze-dried peptides in water. The dialyzers were operated in a countercurrent fashion at a feed flow rate of 12 L/h. The D-6 and BLX 16 GW dialyzers were additionally tested at feed flow rates of 24 and 6 L/h, respectively. Dialysate was provided at a flow rate of 30 L/h from the permeate stream of a reverse osmosis unit (TW-FT30-4021, FilmTec Inc., Minneapolis, MN). Two peristaltic pumps, one placed on either side of the dialyzer, were used to prevent convective mass transfer by matching their flow rates (Figure 1).

A set of tracing experiments using glucose oligomers were performed to test the model presented in the Appendix. Enzyme catalytic behavior was then investigated in both batch experiments and the reactor system to examine the impact of using a continuous reactor on product size distribution and enzyme kinetics.

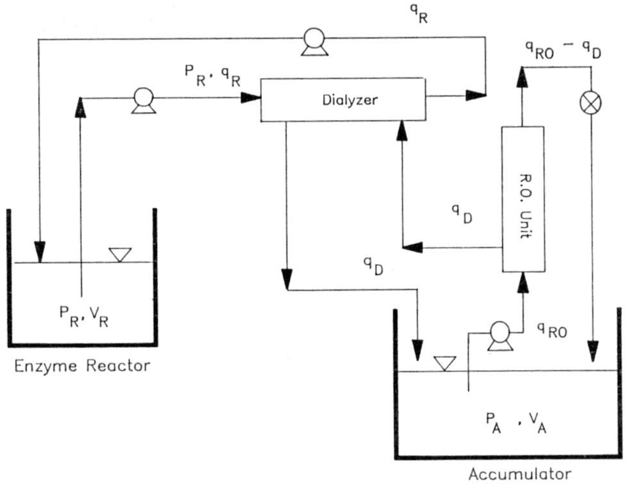

Figure 1. Schematic of the reactor system.

Tracing Experiments

Based on the mass-transfer coefficients estimated from the glucose oligomer studies, experiments were performed to determine the ability of the mathematical models to predict the time course of product accumulation in the accumulator. To test the model in absence of product generation, a solution of glucose oligomers (50 g/L) was pumped through the dialyzers at a flow rate of 12 L/h and a temperature of 50°C. Samples were drawn from the dialyzer inlet and the accumulator every 15 minutes for two hours. Changes in the reactor volume, due to osmotic pressure gradients, were prevented by adjusting the flow rate of the dialyzer outlet pump.

To validate the mathematical model under conditions of constant product generation, similar tracing experiments were performed with product generation being simulated by infusion of a concentrated solution of glucose oligomers (330 g/L) into the reactor at 0.03 L/h.

Batch Experiments

Endopeptidases. A mixture of porcine trypsin and bovine chymotrypsin (Sigma Chemical Co., St. Louis, MO.), each at an enzyme to substrate ratio of 1:100 by weight, was used to hydrolyze whey proteins into peptides. Whey was obtained as a 75% protein powder (Express Foods, Louisville, KY). A 50 g/L solution of the whey proteins was dialyzed against deionized water at pH 3 and pH 7, for two hours each, to remove electrolytes, ash, and other impurities. This dialyzed substrate was reacted, in concentrations of 20, 30, 40, and 50 g protein/L, with the enzyme mixture at 50°C. Samples of the reaction mixture were collected at 5, 10, and 15 seconds after starting the reaction. The samples were quenched with 12 mol/L HCl to stop the enzymatic reaction, then centrifuged at 50000 x g for 20 minutes to separate the soluble peptide species from insoluble high molecular weight proteins. The peptide fragmentation patterns of the supernatants were determined by HPLC using a TSK 2000SW and a TSK 1000PW column connected in series. The mobile phase was acetonitrile and water (40:60), containing 0.096% trifluoroacetic acid, by volume, flowing at 0.036 L/h. Peptide species were detected by UV absorbance at 220 nm. Production rate was determined by subtracting the integrator counts of a control sample from those of the quenched instantaneous samples.

<u>Exoglucosidase.</u> β-amylase from sweet potato (E.C. 3.2.1.2, Sigma, St. Louis, MO), an exoglucosidase capable of cleaving one maltose molecule at a time from the terminal end of 1,4-polysaccharides, was used to produce maltose from soluble starch. Maltodextrin M100 (V-Labs, Covington, LA) was used as the primary substrate source. An aqueous solution of M100 was first diafiltered through a FB-190U UF module (Toyoba, Kyoto, Japan) to obtain a monodispersed substrate with a mean molecular weight of about 35000. Henceforth, this prepared substrate is referred to as M100. The reaction mixture was prepared in a 0.05 mol/L, pH 4.8 acetate buffer containing 0.5 g/L bovine serum albumin. The dependence of the enzyme kinetics on substrate concentration was characterized by performing small volume batch experiments at 37^oC. M100 concentrations were varied from 0.1 to 30 g/L in the presence of 1.2 x 10^3 EU/L enzyme activity. Samples of the reaction mixture were obtained three minutes after initiating the reaction and quenched by immersion in a boiling water bath for five minutes. The samples were then centrifuged for 15 minutes at 50000 x g and the supernatant stored frozen until being assayed. The total sugar concentration in the samples before centrifugation was estimated according to the phenol-sulphuric method (<u>5</u>). The maltose concentration in the supernatant was assayed according to Dygert et al. (<u>6</u>).

<u>Reactor Experiments</u>

Reactor experiments were performed using the same configuration used for the tracing experiments with glucose oligomers (Figure 1)

<u>Endopeptidases.</u> A 50 g/L dialyzed whey protein solution was added to the reactor and brought to 50^oC. The accumulator was initially filled with deionized water. At time zero, the trypsin/chymotrypsin mixture was added to the reactor. Throughout the experiment, the pH of the reactant solution was adjusted with 5 mol/L NaOH to maintain pH 8, the optimal pH for enzyme activity. Samples were regularly drawn from the accumulator. Since the enzymes are retained by the dialyzer membrane, these samples did not need to be quenched. Samples were analyzed for peptides according to the procedure described in the batch experiments section. Reactor and accumulator volumes were kept at 3 L, while dialysate and feed flow rate to the dialyzer were maintained at 30 and 12 L/h, respectively.

<u>Exoglucosidase.</u> The same experimental protocol described for the endopeptidase experiments was followed when β-amylase was used to cleave maltose from the reducing ends of soluble starch molecules. The reaction was run at 37^oC in presence of 39 g/L of M100, and initiated by addition of 0.9 x 10^3 EU/L of enzymatic suspension at time zero. The reaction course was followed for two hours by regularly collecting samples of the reaction mixture. The same buffer and sample handling procedures described for the batch experiments were used. Reactor and accumulator volumes were kept at 1.5 L, while dialysate and feed flow rate to the dialyzer were maintained at 48 and 6 L/h, respectively. No pH adjustment was necessary throughout the reaction because the pH of the reacting solution was monitored and always stayed within the broad pH range over which the enzyme retains its maximum activity (<u>7</u>).

<u>RESULTS</u>

For the tracing experiments, the measured glucose tetramer concentration is compared to that predicted on the basis of the estimated mass transfer coefficients in Figure 2a for the case of zero generation. In Figure 2b,

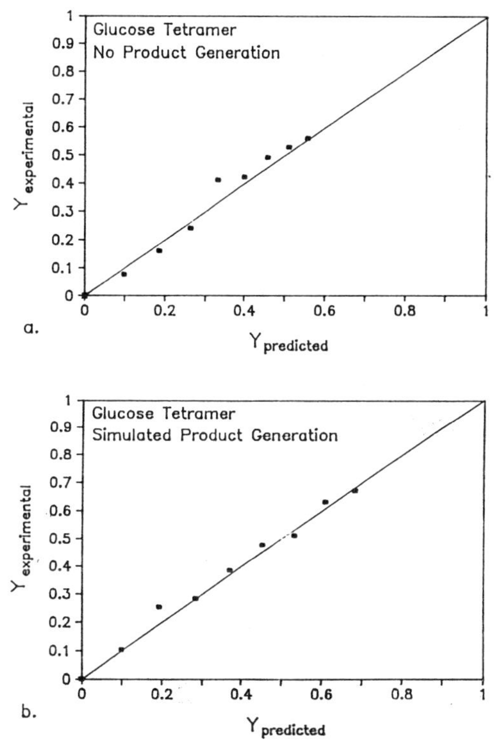

Figure 2. Comparison of measured and predicted fractional concentration for a CF1211L dialyzer.

experimental results and model predictions are compared for the case where product generation is simulated by the continuous addition of a concentrated solution of glucose oligomers to the reactor. Similar results were obtained for the other oligomers.

When using the whey/endopeptidase system, single amino-acids as well as oligopeptides were obtained as reaction products. For ease of mathematical manipulation, integrator counts obtained for the heterogeneous mixture of peptide products were summed over elution times corresponding to 500 molecular weight segments and referred to as a single class of products.

Results of the batch experiments with the endopeptidase mixture are shown in Figure 3. The integrator counts for each molecular weight group increased with time within the first five seconds, and then leveled off. The absolute value of the counts for each molecular weight segment decreased with increasing molecular weight, except for peptides in the 1000-1500 molecular weight range which consistently exhibited greater counts than the other groups. In Figure 4, the dependence of reaction rate (as determined from batch experiments at five seconds into the reaction) on substrate concentration is expressed in terms of a Lineweaver-Burk plot. Based on the linearity of the experimental data, the enzyme is assumed to exhibit Michaelian kinetic behavior. Values for the reaction rate at saturation, V_{max}, and the Michaelis constant, K_M, for each molecular weight group were estimated from a non-linear fitting procedure and are reported in Table 1.

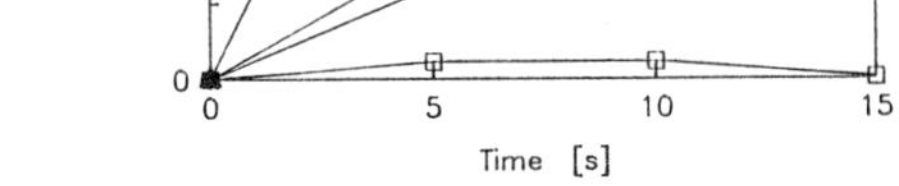

Figure 3. Product formation for a 50 g/L whey protein solution in a batch system. Data refer to different molecular weight ranges: (□) 0-500, (●) 500-1000, (Δ) 1000-1500, (■) 1500-2000.

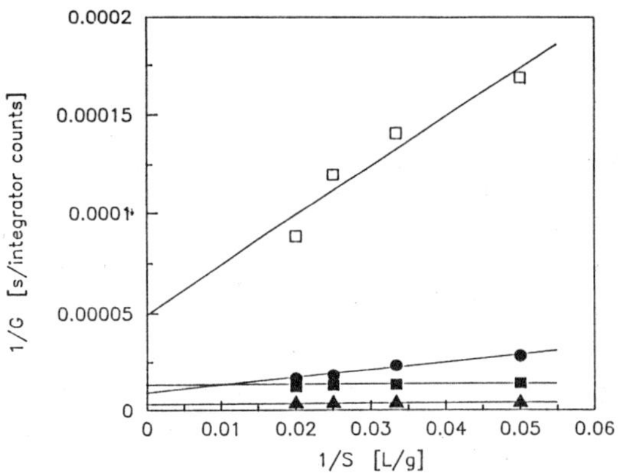

Figure 4. Lineweaver-Burk plot for the whey/endopeptidases system. Data are reported for products of different molecular weight: (□) 0-500, (●) 500-1000, (Δ) 1000-1500, (■) 1500-2000.

When the endopeptidases were used in the closed-loop reactor system, product appearance in the accumulator was monitored serially. The time course of product concentration in the accumulator is shown in Figure 5 for different molecular weight groups and operating conditions. Product concentration in the accumulator steadily increased with time, with concentrations of peptides within the 1000-1500 and the 500-1000 molecular weight ranges being greater than those of other groups. Concentrations in the 500-1000 molecular weight range were consistently greater than those in the 1000-1500 range. Greater absolute product concentrations were obtained with the D-6 dialyzer. The typical time courses of molecular weight distribution of the product in the accumulator are shown in Figure 6, where integrator counts are reported against molecular weight at different times into the reaction. Similar results and time patterns

Table 1. Kinetic parameters for the whey/endopeptidases system from batch experiments. K_M is referred to the primary protein component in whey, the β-lactoglobulin dimer (M.W. 34000).

Molecular Weight Range	$G^0 \times 10^{-8}$ Integrator units/h	$K_M \times 10^6$ mol/L
0- 500	0.73	1471
500-1000	4.21	1394
1000-1500	11.04	82
1500-2000	3.19	143

were obtained at other flow rates and with other dialyzers. Although product concentrations steadily increased, the molecular weight distribution was determined after 15 minutes into the reaction and remained constant over the observed period of time.

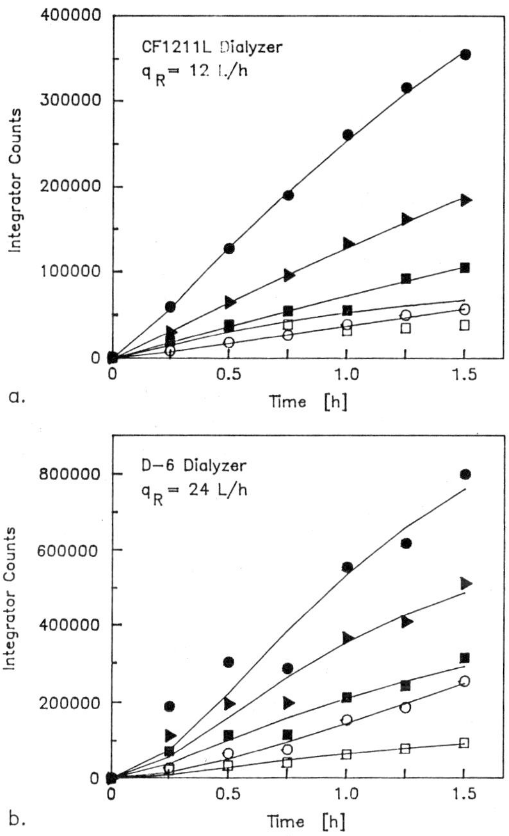

Figure 5. Concentration of peptide fractions in the accumulator as a function of time. Lines represent the best fit of the model to the data and the symbols represent experimental data for different molecular weight ranges: (□) 0-500, (●) 500-1000, (Δ) 1000-1500, (■) 1500-2000, ( ) 3500-4000.

The kinetics of the cleavage of M100 by β-amylase were investigated in a batch system according to an initial rate technique, where all reaction rates were estimated after three minutes. The enzyme exhibited Michaelian kinetics, with a Michaelis constant, K_M, for the cleavable fraction of the M100 of 1.1 g/L. Competitive inhibition kinetics were assumed for β-amylase (7). Fitting of the theoretical model to experimental maltose concentrations in the reactor led to an optimal value for the inhibition constant of 1.87 g/L of maltose. Predictions from the theoretical model matched closely experimental data in the reactor and in the accumulator until about 30 minutes, Figure 7. Beyond that point, the model slightly overpredicted maltose concentrations and, therefore, the reactor yield.

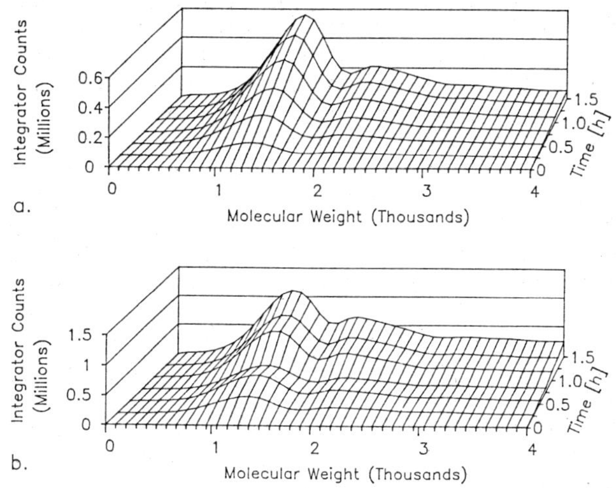

Figure 6. Change with time of peptide product concentration and molecular weight distribution in the accumulator.
a. Data for a CF1211L dialyzer at a feed flow rate of 12 L/h.
b. Data for a D-6 dialyzer at a feed flow rate of 24 L/h.

DISCUSSION

The kinetic behavior of endopeptidases, such as the ones used in this investigation, is known to be affected to a great extent by reaction products (8). Even though little is understood of the mechanism of enzyme action, the products of the random cleavage of long polypeptides certainly act as strong

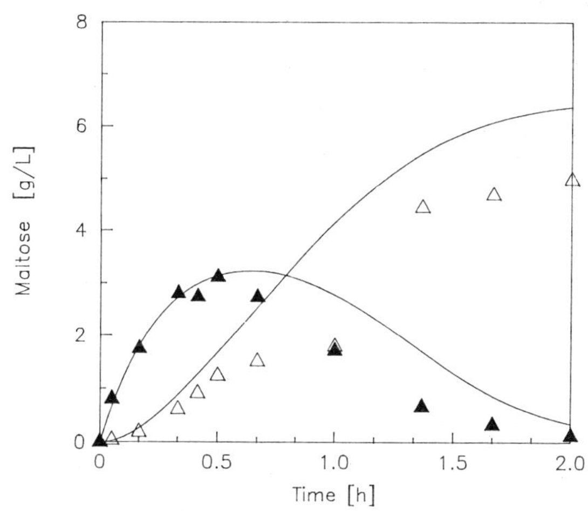

Figure 7. Time course of maltose accumulation in the closed-loop reactor. Concentrations in the reactor (Δ) and in the accumulator (Δ) are compared to model predictions, reported as solid lines.

inhibitors of enzymatic catalysis. As shown in Figure 3, when enzyme kinetic behavior is tested in a batch system, build up of a small amount of products prevents further substrate conversion from occurring and the product generation rate is virtually reduced to zero. In the case of either random or sequential attack on the polymeric chain, the kinetics of enzymatic substrate cleavage can still be modelled in terms of the Michaelis-Menten rate equation (9), thus producing linear Lineweaver-Burk plots (Figure 4). Since the substrate concentration in the reaction mixture was generally greater than the enzyme Michaelis constant (Table 1), it was assumed that peptide generation rate is independent of substrate concentration and depends on the molecular weight group only.

In the past, systems have been proposed to overcome this sudden and highly detrimental feedback inhibition. Confinement of the enzyme/substrate mixture in dialysis bags was proposed as a suitable way to carry out the enzymatic process as early as 1955 (10). Although based on the same principle, the closed-loop reactor system investigated in this paper also allows the simultaneous recovery of concentrated low molecular weight products. System fluid dynamics were modelled according to the integral mass balance equations presented in the Appendix. Tracing of the reactor system with a mixture of glucose oligomers fully validated the mathematical approach, both in the case of zero generation rate and with a constant generation rate. The good agreement between experimental and predicted sugar concentrations in the accumulator is shown in Figure 2.

When the initial generation rate, as determined by the Lineweaver-Burk plot from the batch experiments, was used to predict peptide appearance rates in the accumulator following enzymatic cleavage of whey proteins, the measured concentrations were far greater than those predicted by the model (Figure 8). These findings imply that the value of G^0 can substantially exceed that estimated from the batch experiments.

By allowing for thermal and mechanical denaturation of the enzyme (see Appendix), good agreement between the experimental and theoretical concentration profiles could be obtained for the reactor experiments by curve fitting (Figure 5). Values for the initial enzymatic activity, G^0, and the denaturation constant, γ, obtained from a non-linear

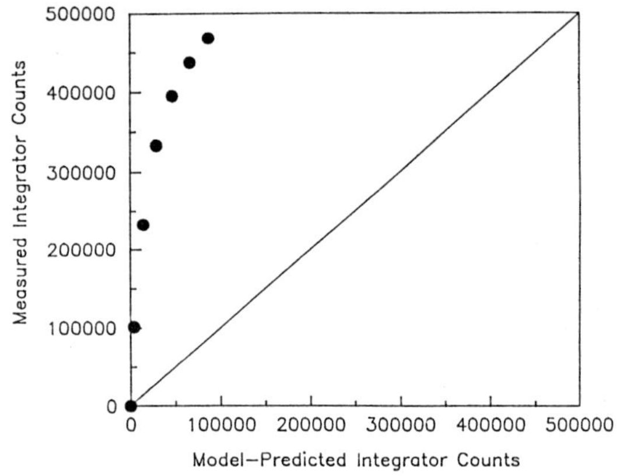

Figure 8. Measured and predicted peptide product concentrations when using a D-6 dialyzer at a feed flow rate of 12 L/h for separating peptides in the 500-1000 molecular weight range. Model predictions are based on batch generation rate.

regression to the experimental data, are reported in Table 2 for each molecular weight segment. Continuous product removal from the reactor by dialysis resulted in enzyme reaction rates up to three order of magnitudes greater than when the reaction was performed in a batch system. The kinetic performance of the system at different product extraction rates was established by using dialyzers with a low permeability (CF1211L) or a high permeability (D-6) membrane, or by changing the feed flow rate to a given dialyzer. For each molecular weight group, faster product extraction rates generally resulted in higher product concentrations in the accumulator, at any given time (Figure 5). The enhancement of reactor productivity can be attributed to both the increase in mass transfer through the dialysis membrane and the resulting relief from product inhibition.

Table 2. Comparison of G^0, [integrator counts/h], and γ, [h^{-1}], for the three operating conditions.

Dialyzer	CF1211L		D-6			
Feed Flow Rate	12 L/h		12 L/h		24 L/h	
M.W. Range	$G^0 \times 10^{-10}$	γ	$G^0 \times 10^{-10}$	γ	$G^0 \times 10^{-10}$	γ
0- 500	2.2	55.5	0.7	25.3	0.05	0.96
500-1000	6.2	15.1	1.9	7.9	0.6	1.32
1000-1500	24.5	46.0	7.1	35.8	1.1	3.54
1500-2000	11.5	35.8	6.2	37.4	1.5	6.36
2000-2500	7.8	21.5	11.7	45.7	1.3	3.54
2500-3000	16.6	27.6	36.2	37.4	1.7	2.94
3000-3500	11.3	15.7	65.4	60.0	1.1	2.28
3500-4000	13.0	12.8	180	56.5	1.5	2.16

A comparison between Figures 3 and 5 shows how strongly continuous product removal affects enzyme action. Not only does continuous product removal result in higher yields, but the protein fragmentation pattern is changed, as well. Intermediate products, which tend to accumulate on their way to total degradation to amino acids, feature a molecular weight in the 500-1000 range rather than the 1000-1500 range observed for the batch system. Because of product removal, substrate degradation is thus accomplished a step further. These results suggest that, by careful selection of the dialyzer and the system fluid dynamics, the average molecular weight of the final products can be maintained within a desired range. In Figure 6 and Table 3, molecular weight distribution and number average molecular weight are reported for different dialyzers and feed flow rates. The lowest molecular weight products were obtained when the CF1211L dialyzer was used for product removal, while the molecular weight distribution was slanted towards the higher molecular weight species when the D-6 dialyzer was used. Provided that low molecular weight products are removed rapidly enough to prevent their accumulation in the reactor, the longer the substrate molecules are kept in contact with the enzyme, the more susceptible they are to further cleavage. As a result, the average molecular weight of the products in the accumulator will be lower as the dialyzer becomes less permeable.

Table 3. Mean molecular weight of peptide products in the accumulator for the three operating conditions.

Dialyzer	CF1211L	D-6	
Feed Flow Rate	12 L/h	12 L/h	24 L/h
Molecular Weight			
Number Average	755.6	768.0	870.5
Weight Average	1183.1	1402.4	1660.0
Mean Equivalent Weight	353.0	418.0	463.0

As far as the β-amylase/starch system is concerned, the value of the inhibition constant for maltose, estimated to be 1.87 g/L, is in good agreement with the value of 2 g/L reported by Misra et al. (11). The theoretical model (see Appendix) accurately predicts maltose concentrations in the accumulator and the reactor for the first 30 minutes, after which product concentrations are overpredicted (Figure 7). This discrepancy may result either from thermal and mechanical enzyme denaturation, as is the case of the protease system, or from substrate loss into the dialysate. As compared to natural substrates, such as cheese whey, the diafiltered M100 features a narrower molecular weight distribution that makes it more likely to experience a reduction of average molecular weight under the attack of hydrolytic enzymes (12). Indeed, substrate concentration in the reactor decreased more rapidly than expected based on maltose production.

Contrary to proteases, β-amylase is inhibited by reaction products only at higher concentration levels, as the inhibition constant for maltose suggests. Under these circumstances, two competing effects will determine the final reactor performance: on one hand the positive effect of product removal on enzyme action, and on the other hand, the progressive loss of substrate by dialysis once its molecular weight has been sufficiently reduced to allow it to permeate the dialysis membrane. Further investigation is required to establish the relative contribution of these two effects, as well as to elucidate the mechanism of enzyme catalytic action.

CONCLUSIONS

The use of membranes coupled to an enzyme reactor to relieve the effect of product inhibition has generally involved the use of UF (13,14). UF is used either in closed or open loop with an enzyme reactor, and high permeate fluxes are required to obtain high products recovery. This often results in poor reactor performance after a short operation time because of concentration polarization and membrane fouling (15,16). Since dialysis is a transport process driven by a concentration gradient, no net permeate flux is necessary for mass transfer. Provided that proper fluid dynamic conditions are established on both sides of the membrane, the adverse effects of fouling and, to a lesser extent, concentration polarization can be reduced, if not eliminated. Moreover, combination of the reactor/ dialyzer system with a RO unit in a closed-loop configuration allows continuous regeneration of the dialysate and concentration of the low molecular weight product. The system proposed in this investigation provides a feasible way to perform product inhibited enzymatic reactions in an efficient way. With strongly inhibited reactions, improvements over traditional batch techniques are remarkable. However, for reactions suffering from product inhibition only at high product concentrations, the use of the system is advantageous only if concentrated, natural substrates are used. Once the enzymatic system is chosen, selection of appropriate fluid dynamic conditions will result in optimal performance of the entire system and allow the characteristics of the final products to be controlled.

ACKNOWLEDGEMENTS

This work was supported in part with funds from the National Science Foundation (Grant No. RII-8610671) and the Commonwealth of Kentucky through the Kentucky EPSCoR Program.

NOTATION

Roman Letters

A = membrane surface area [m^2]
G = reaction rate [integrator counts/s]
K_I = inhibition constant [g/L]
K_M = Michaelis constant [g/L]
N_T = $K\ A/q_R$
P = product concentration [g/L]
S = substrate concentration [g/L]
Y = P_A/S_R
V_{max} = reaction rate at saturation [g/L/s]
Z = q_R/q_D

Greek Letters

α = $q_R\ (1 - \beta)/(\beta\ V_R)$
β = $(e^{N_T(1-Z)} - Z)/(1-Z)$
γ = deactivation constant [h^{-1}]

Subscripts and Superscripts

A,R = accumulator and reactor
D,RO = dialyzer and reverse osmosis unit
o = initial value

LITERATURE CITED

1. Klein, E., R.A. Ward, R.E. Lacey, "Membrane Processes -Dialysis and Electrodialysis", in Handbook of Separation Process Technology, Rousseau, R.W. (Ed.), J. Wiley & Sons, New York (1987).

2. Perry, J.H. and C. Chilton, Chemical Engineers' Handbook, 5th ed., McGraw-Hill, New York (1973).

3. Klein, E., R.A. Ward, T.E. Williams, P.W. Feldhoff, Trans. Am. Soc. Artif. Int. Organs, XXII, 550 (1986).

4. Klein, E., T.J. Turnham, R.A. Ward, Artif. Organs, 12 (3), 252 (1988).

5. Dubois, M., K.A. Gilles, J.K. Hamilton, P.A. Rebers, F. Smith, Anal. Chem., 28, 350 (1956).

6. Dygert, S.L., L.H. Li, D. Florida, J.A. Thoma, Anal. Biochem., 13, 367 (1965).

7. French, Dexter, "β-amylases", in The Enzymes, Vol. 4, 2nd ed., Boyer, P.D., H. Lardy, K. Myrbach (Eds.), Academic Press, New York (1960).

8. Gauthier, S.F., C. Vachon, J.D. Jones, L. Savoie, J. Nutr., 112, 1718 (1982).

9. Thoma, J.A., J.E. Sprandlin, S. Dygert, "Plant and Animal Amylases," in The Enzymes, Vol. 5, Boyer, P.D. (Ed.), Academic Press, New York (1971).

10. Mauron, J., F. Mottu, E. Bujard, R.H. Egli, Arch. Biochem. Biophys., 58, 433 (1955).

11. Misra, U.K. and D. French, "Inhibition and Action Pattern of β-amylase in Presence of Maltose", in Biochem. J., Proc. of the Biochem. Soc., 77, P1 (1960).

12. Thoma, John A., Biochemistry, 5 (4), 1365 (1966).

13. Henley, R.G., R.Y. Yang, P.F. Greenfield, Enzyme Microb. Technol., 2, 206 (1980).

14. Deeslie, W.D. and M. Cheryan, Biotechnol. Bioeng., XXXIII, 2257 (1981).

15. Closset, G.P., J.T. Cobb, Y.T. Shah, Biotechnol. Bioeng., XVI, 345 (1974).

16. Cheryan, M. and W.D. Deeslie, "Production of Protein Hydrolyzates in Ultrafiltration-Enzyme Reactors," in Ultrafiltration Membranes and Applications, Cooper, A.R. (Ed.), Plenum Press, New York (1980).

17. Williams, Terry E., "Preparation of Peptides for Use as Osmotic Agents in

Peritoneal Dialysis," PhD thesis, Univ. of Louisville, Louisville (1988).

18. Alfani, F., D. Albanesi, M. Cantarella, V. Scardi, Enzyme Microb. Technol., 4, 181 (1982).

19. Greco, G., Jr., L. Gianfreda, D. Albanesi, M. Cantarella, J. Appl. Biochem., 3, 233 (1981).

20. Robbins, R.C., Internat. J. Vit. Nutr. Res., 48, 44 (1978).

APPENDIX

The concentration of any species collected in the accumulator may be described as a function of time through integral mass balances about the various components in the separation process (17). In the development of the model, the following assumptions were made:

i. reactor and accumulator are well mixed;
ii. the dynamics of both membrane units, i.e. RO and dialyzer, are negligible compared to that of the reactor and the accumulator;
iii. a species' mass-transfer coefficient is independent of time.

Mass balances about the reactor and the accumulator can then be written as follows:

$$\frac{d\,(V_R\,P_R)}{dt} = q_R\,(1/\beta - 1)\,P_R + G(t,P_R) \qquad (1)$$

$$\frac{d\,(V_A\,P_A)}{dt} = q_R\,(1 - 1/\beta)\,P_R \qquad (2)$$

with the following initial conditions:

$t=0 \quad P_A = 0\ ;\ P_R = P_R^{o}$.

Constant Product Generation Rate. When substrate is in large excess, and reaction inhibition is minimized by continuous product withdrawal, the product generation rate can be assumed constant. Upon integrations of Equations (1) and (2), product concentration in the accumulator can be expressed as follows:

$$P_A = \left[\frac{V_R}{V_A}P_R^{o} + \frac{G}{\alpha\,V_A}\right](1 - e^{\alpha t}) + \frac{G\,t}{V_A} \qquad (3)$$

In the absence of generation, i.e. for G=0, Equation (3) becomes:

$$P_A = \frac{V_R}{V_A}P_R^{o}\,(1 - e^{\alpha t}) \qquad (4)$$

Exponential Decay of Product Generation. A decreased product generation with time can be expected to occur in an enzyme system where the catalyst is susceptible to mechanical (18) and thermal (19) degradation as well as autolysis (20). The loss of enzyme activity was modelled exponentially with time according to an Arrhenius relationship (19):

$$G(t) = G^{o}\,e^{(-\gamma\,t)} \qquad (5)$$

By substituting Equation (5) into Equation (1) and integrating, the product concentration in the accumulator tank can be expressed according to:

$$P_A = \frac{Q_R}{V_A}\left(1 - \frac{1}{\beta}\right)\left(\frac{G^{o}/V_R}{\gamma + \alpha}\right)\left[\frac{1}{\alpha}(e^{\alpha t} - 1) + \frac{1}{\gamma}(e^{-\gamma t} - 1)\right] \qquad (6)$$

Competitive Inhibition. β-amylase has been reported to be competitively inhibited by the maltose cleaved from long polysaccharide molecules (11). This kinetic behavior was described according to the following equation:

$$G(P_R,\,S_R) = \frac{V_{max}\,S_R}{K_M\left(1 + \frac{P_R}{K_I}\right) + S_R}\,V_R \qquad (7)$$

By substitution of Equation (7) into Equation (1), the resulting set of mass balance partial differential equations was numerically fitted to experimental data to give the optimal value of K_I.

DEVELOPMENT OF A NOVEL OXIDATIVE PALLADIUM MEMBRANE REACTOR

N. Itoh and R. Govind ■ Department of Chemical Engineering, University of Cincinnati, Cincinnati, OH 45221

The use of palladium and its alloys in a membrane reactor is attractive because the material is permeable only to hydrogen, can withstand high temperatures, and exhibits catalytic properties for hydrogenation and oxidation reactions. Furthermore, palladium being a good heat conductor, allows effective heat transfer to and from the reactor. One of the disadvantages of a palladium membrane is its low permeability for hydrogen.

In this paper, simulation results have been presented for a palladium membrane reactor for the dehydrogenation of 1-butene with the permeating hydrogen being oxidized to water by oxygen on the permeation side.

INTRODUCTION

Recent studies on membrane reactors have identified its potential advantages over conventional tubular and stirred tank reacting systems [Mohan and Govind, 1988]. However, implementing the concept of reaction with simultaneous separation is difficult, due to the unavailability of membranes which can withstand high temperatures and exhibit high permeabilities and selectivities. Furthermore, heat addition to or removal from the reactor, to maintain the reactor temperature, is essential especially if high conversions are attained in the membrane reactor. The existence of a permeation or separation side in a membrane reactor makes it difficult to remove or add heat to the reactor.

In this paper a theoretical simulation study has been presented for a counter- current flow palladium membrane reactor system with an oxidation reaction on the permeation side and dehydrogenation of 1-butene on the reaction side of the membrane. The main advantages of using palladium and its alloy membranes are:

1. The membranes can withstand high temperatures.

2. The membranes are highly selective to the permeation of hydrogen.

3. The membranes exhibit catalytic activity for oxidation and dehydrogenation reactions.

4. The membranes can effectively transfer heat from an exothermic oxidation reaction on the permeation side to an endothermic dehydrogenation reaction on the other side (reaction side) of the membrane.

5. The membranes can be made very thin for achieving high hydrogen fluxes.

BACKGROUND

Using membranes to shift the equilibrium in a reversible reaction has been recognized since the late sixties (Wood 1968; Shah et al. 1970; Barker and Burns, 1973; Gryaznov et al. 1973; Hwang and Kammermeyer, 1975). Practical application has been limited to inorganic membranes, due to the lack of suitable membranes that can withstand high temperatures and provide good permselectivities.

In recent years, many researchers have studied the feasibility of producing hydrogen by enhancing the conversion of dehydrogenation/decomposition reactions. Dokiya et al. (1977), Fukuda et al. (1978), and Kameyama et al. (1981), employed porous Vycor glass to selectively separate the hydrogen produced in the decomposition of hydrogen sulphide. In a separate study using the same reaction, Kameyama et al. (1983) showed the possibility of using porous alumina membranes. Shindo et al. (1981) conducted similar studies for the hydrogen iodide reaction using porous Vycor glass. Itoh et al. (1984), in a theoretical study of the decomposition of hydrogen iodide, showed that significant equilibrium shift is possible when porous Vycor glass is used to separate the hydrogen. Shinji et al. (1982) and Itoh(1987), using porous Vycor glass and palladium respectively, enhanced the dehydrogenation of cyclohexane. Palladium being permeable only to the hydrogen, high conversions could be achieved by increasing residence time; i.e., decreasing the feed flow rate. Itoh et al. (1985) simulated the membrane reactor with the above reaction and showed that for given rates of permeation and reaction, there is an optimum thickness of membrane at which maximum conversion is obtained. In a separate parametric study (Mohan and Govind, 1986, 1988a), it was shown that for a membrane which is permeable to both products and reactants, the maximum equilibrium shift possible is limited by the loss of reactant, due to permeation, from the reaction zone. However, for membranes which have low (or zero) permselectivity for the reactant, a tremendous reduction in reactor volume (over impermeable wall reactor) can be made while achieving significant (or complete) conversion. Furthermore, in the case of dehydrogenation reactions with membranes that only permeate hydrogen, conversions comparable to those achieved with lesser permselective membranes can be attained at substantially lower feed temperatures (Mohan and Govind, 1988b).

The coupling of hydrogenation and dehydrogenation reactions in a catalytic membrane reactor was first proposed by Gryaznov et al. (1973, 1986) and Itoh (1988) though no quantitative treatment of the concept was presented. Recently, a theoretical study of coupling oxidation of hydrogen with dehydrogenation of 1- butene using an isothermal/adiabatic co-current palladium membrane reactor system has been presented (Itoh and Govind, 1988).

MODEL DEVELOPMENT

A schematic of the palladium membrane reactor is shown in Figure 1, with oxidation of hydrogen by air to water on the separation side (palladium membrane as catalyst) and catalytic dehydrogenation of 1-butene to butadiene on the reaction side. The catalyst is packed uniformly on the reaction side. The hydrogen produced by the dehydrogenation reaction permeates through the membrane and reacts with the oxygen on the separation side. This surface reaction of hydrogen with oxygen decreases the hydrogen concentration on the separation side, thereby increasing the permeation of hydrogen through the membrane. Further, heat liberated by the exothermic oxidation reaction on the separation side flows across the membrane and facilitates the endothermic dehydrogenation reaction on the reaction side.

The essential assumptions for the development of the model equations summarized in Table 1 were as follows:

1. Plug flow on both sides of the membrane due to high Reynolds numbers typically found in flow through packing or flow through small diameter tubes.

2. No axial or radial diffusion, due to dominance of convective flow over diffusive flow (high Peclet numbers).

3. Insignificant pressure drop on either side of the membrane, since it is a small fraction of the total pressure.

The parameters used in the simulation study have been summarized in Table 2. The kinetic model for the dehydrogenation of 1-butene and the surface reaction between hydrogen and oxygen have been presented by Happel et al. (1966) and by Leder and Butt (1966), respectively.

RESULTS AND DISCUSSION

Two modes of operation, co-current and counter-current were simulated using isothermal conditions. For the co-current case, the hydrogen concentration profile on the reaction and separation side is shown in Figure 2. On the reaction side, there is a rapid increase in hydrogen concentration due to reaction followed by a decline due to permeation of hydrogen through the membrane. Eventually a dynamic equilibrium is established when the hydrogen generation rate on the reaction side is balanced by the hydrogen permeation rate to the separation side. On the separation side, there is an increase in the hydrogen concentration due to permeation from the reaction side. It is important to note, that for the operating parameters selected in Figure 2, the state of dynamic equilibrium is achieved when the hydrogen concentration on the reaction and separation sides are equal, i.e., no permeation of hydrogen through the membrane. This is due to the fact that in this case, the hydrogen rate on the reaction side is much greater than the rate of hydrogen consumption on the separation side due to oxidation.

For the counter-current case, as shown in Figure 3, the hydrogen concentration increases rapidly at the entrance of the reactor due to the dehydrogenation reaction. After this rapid increase, the concentration of hydrogen steadily declines due to permeation to the separation side. On the separation side, the hydrogen concentration continues to increase along the reactor length beginning at zero concentration at the reactor exit. Note that for counter-current operation, the flow on the separation side is opposite in direction to the flowrate on the reaction side. Furthermore, in this case, there exists a concentration difference between the hydrogen concentrations on the reaction and separation sides along the entire reactor length.

The conversion profiles for the co-current and counter-current cases is shown in Figure 4. For the co-current case, the conversion initially increases but then levels at a certain value when the hydrogen concentrations on the reaction and separation sides become equal. The maximum conversion attained increases when the hydrogen permeability is increased but attains an asymptotic value which is governed by the flowrates of the reactant on the reaction side, inerts on the separation side, and other operating parameters such as pressure on the reaction and separation sides and temperature. For the counter-current case, the conversion also increases as for the co-current case, but the conversions attained for the counter-current case are significantly higher than the co- current system for the same hydrogen permeation rate. The difference between the conversions attained in the counter-current and co-current membrane reactor systems increases with increasing hydrogen permeation rate. Higher conversions are attained with the counter-current mode due to higher hydrogen fluxes through the membrane caused by a greater concentration difference between the reaction and separation sides.

CONCLUSIONS

It has been shown through computer simulation that coupling an exothermic hydrogen oxidation reaction with dehydrogenation of 1-butene in an isothermal palladium membrane reactor results in higher conversions than can be achieved in a tubular reactor. Furthermore, counter-current mode of operation gives a much higher conversion than the co-current mode for the same hydrogen permeability through the membrane. Finally, palladium is a suitable membrane material for separating hydrogen from reacting systems, since it permeates only hydrogen, thereby eliminating reactant loss to the separation side.

NOMENCLATURE

A = membrane area, m^2

$c^0_{P_i}$ = specific heat of reactant, $J.mol^{-1}.K^{-1}$

c_{P_i} = specific heat of gas i at a temperature T, $J.mol^{-1}.K^{-1}$

C_{P_i} = $c_{P_i}/c^0_{P_i}$

C_0 = concentration of hydrogen dissolved in palladium membrane at T_0 and P_0, $mol.m^{-3}$

D = diffusion coefficient of hydrogen, $m^2.s^{-1}$

Da^0_r = Damkohr number for reaction side at T_0, $k_r G P_r^{1/2}/u^0_c$

Da^0_s = Damkohr number for reaction side at T_0, $k_s A P^3_s/u^0_c$

DI_r = dilution ratio, u^0_I/u^0_c

DI_s = dilution ratio, v^0_I/u^0_c

E = activation energy, $J.mol^{-1}$

G = total weight of catalyst, g

h_0 = overall heat transfer coefficient, $J.m^{-2}.s^{-1}.K^{-1}$

ΔH = heat of reaction, $J.mol^{-1}$

k_r = rate constant of dehydrogenation, $mol.g - cat.^{-1}.s^{-1}.Pa^{-1/2}$

k_s = rate constant of oxidation, $mol.m^{-2}.s^{-1}.Pa^{-3}$

K = equilibrium constant for dehydrogenation of 1-butene, Pa

L = dimensionless reactor length

m = stoichiometric mole number of hydrogen produced in dehydrogenation

p_i = partial pressure of gas i, Pa

P_r = total pressure on reaction side, Pa

P_s = total pressure on separation side, Pa

P_0 = reference pressure, 1.01325×10^5 Pa

R = gas constant, $J.mol^{-1}.K^{-1}$

t_m = thickness of membrane, m

T = absolute temperature, K

Tu^0 = dimensionless number relating hydrogen permeation at T_0 and P_0, $DC_0A/t_m u_c^0$

T_0 = absolute temperature at inlet of reactor, K

u_i = of gas i in reaction-side stream, $mol.s^{-1}$

u_i^0 = of gas i in reaction-side inlet, $mol.s^{-1}$

U_i = dimensionless of gas i in reaction-side stream, u_i/u_C^0, u_H/mu_C^0 for U_H

v_j = of gas j at separation-side stream, $mol.s^{-1}$

v_j^0 = of gas j at separation-side inlet, $mol.s^{-1}$

V_j = dimensionless of gas j in separation-side stream, v_j/u_C^0, v_H/mu_C^0 for V_H

γ = $h_0A/c_{P_c}^0 u_c^0$

δ = $\Delta H/c_{P_c}^0 T_0$

ε = E/RT_0

θ = T/T_o

= Subscript

C = reactant of dehydrogenation

D = dehydrogenation product

H = hydrogen

i = component i in reaction-side stream

I = inert gas

j = component j in separation-side stream

o = oxygen

p = permeation

r = reaction side

s = separation side

W = water

= Superscripts

0 = indicate value at the inlet of reactor

LITERATURE CITED

Barker, S.A., and R.F. Burns, "Reactor Separators Incorporating Membrane Bound Enzymes," *Chemistry and Industry*, (Lond.),16, 801 (1973).

Dokiya, M., T. Kameyama, and K. Fukuda, "The Application of the Effusion on the Thermochemically Limited Reaction," *Denki Kagaku*, 45, 701 (1977).

Fukuda, K., M. Dokiya, T. Kameyama, and Y. Kotera, "Catalytic Decomposition of Hydrogen Sulfide," *Ind. Eng. Chem. Fundam.*, 17, 243 (1978).

Gryaznov, V.M., V.S. Smirnov and G. Slinko, "Heterogeneous Catalysis with Reagent Transfer through the Selectively Permeable Catalysts," *Catalysis*, (J.W. Hightower, ed.), American Elsevier, N.Y. (1973).

Gryaznov, V.M., "Hydrogen Permeable Palladium Membrane Catalysts," *Platinum Metals Rev.*, 30, 68 (1986).

Happel, J., H. Blank, T.D. Hamill, *I&EC Fundamentals*, 5, 289 (1966).

Hwang, S.T., and K. Kammermeyer, *Membranes in Separations*, Wiley-Interscience, N.Y. (1975).

Itoh, N., Y. Shindo, T. Hakuta and Y. Yoshitome, "Enhanced Catalytic Decomposition of HI by Using a Microporous Membrane," *Int. J. Hydrogen Energy*, 9, 835 (1984).

Itoh, N., Y. Shindo, K. Haraya, K. Obata, T. Hakuta and H. Yoshitome, "Simulation of a Reaction

Accompanied by Separation," *Int. Chem. Eng.*, **25**, 138 (1985).

Itoh, N., "A Membrane Reactor Using Palladium," *A.I.Ch.E. Journal*, **33**, 1576 (1987).

Itoh, N., personal communication (1988).

Itoh, N., and R. Govind, "Combined Oxidation and Dehydrogenation in a Palladium Membrane Reactor," paper submitted to I&EC Research, (1988).

Kameyama, T., M. Dokiya, M. Fujishige and H. Yokokawa, "Possibility for Effective Production of Hydrogen from Hydrogen Sulfide by means of a Porous Vycor Glass Membrane," *Ind. Eng. Chem. Fund.*, **20**, 97 (1981).

Kameyama, T., M. Dokiya, M. Fujishige, H. Yokokawa and K. Fukuda, "Production of Hydrogen from Hydrogen Sulfide by means of Selective Diffusion Membranes," *Int. J. Hydrogen Energy*, **8**, 5 (1983).

Leder, H., and J.B. Butt, "Surface Catalysis of the Hydrogen-oxygen Reaction at Low Temperatures," *A.I.Ch.E. Journal*, **12**, 718 (1966).

Mohan, K., and R. Govind, "Analysis of a Cocurrent Membrane Reactor," *A.I.Ch.E. Journal*, **32**, 2083 (1986).

Mohan, K., and R. Govind, "Analysis of Equilibrium Shift in Isothermal Reactors with a Permselective Wall," *A.I.Ch.E. Journal*, **34**, 1493 (1988a).

Mohan, K., and R. Govind, "Effect of Temperature on Equilibrium Shift in Reactors with Permselective Wall," *I&EC Research*, **27**, 2064 (1988b).

Shah, Y.T., T. Rammen and S.H. Chiang, "A Note on Isothermal Permeable Wall Plug Flow Reactor," *Chem. Eng. Sci.*, **25**, 1947 (1970).

Shindo, Y., K. Haraya, K. Obata, T. Hakuta and H. Yoshitome, Preprints for 15th Autumn Meeting Soc. Chem. Eng., Kanazawa, Japan (1981), as referred to by Itoh et al. (1984,1985).

Shinji, O., M. Misonou and Y. Oneda, "The Dehydrogenation of Cyclohexane by the Use of a Porous-glass Reactor," *Bull. Chem. Soc. Japan*, **55**, 2760 (1982).

Wood, B.J., "Dehydrogenation of Cyclohexane on a Hydrogen-Porous Membrane," *J. Catalysis*, **11**, 30 (1968).

Table 1: Basic equations developed for the membrane reactor system

1. Material Balance

Reaction Side

$$\frac{dU_C}{dL} = -Da_r^0 \mathrm{Exp}\left\{\varepsilon_r\left(1-\frac{1}{\theta_r}\right)\right\} f_r \quad (1)$$

$$\frac{dU_H}{dL} = -Da_r^0 \mathrm{Exp}\left\{\varepsilon_r\left(1-\frac{1}{\theta_r}\right)\right\} f_r - \frac{Tu^0}{m}\mathrm{Exp}\left\{\varepsilon_p\left(1-\frac{1}{\theta_r}\right)\right\}\left(\sqrt{\frac{p_{H_r}}{P_0}}-\sqrt{\frac{p_{H_s}}{P_0}}\right) \quad (2)$$

$$U_D = 1 - U_C \quad (3)$$

$$U_r = DI_r \quad (\text{constant for inert gas}) \quad (4)$$

Separation Side

$$\frac{dV_o}{dL} = -Da_s^0 \mathrm{Exp}\left\{\varepsilon_s\left(1-\frac{1}{\theta_s}\right)\right\} f_s \quad (5)$$

$$V_w = 2(V_o^0 - V_o) \quad (6)$$

$$V_H = 1 - U_C - U_H - 2(V_o^0 - V_o)/m \quad (7)$$

$$V_I = DI_s \quad (\text{constant for inert gas}) \quad (8)$$

2. Heat Balance

Reaction Side

$$\frac{d(\sum C_{P_i} U_i \theta_r)}{dL} = -Da_r^0 \mathrm{Exp}\left\{\varepsilon_r\left(1-\frac{1}{\theta_r}\right)\right\}\delta_r f_r + \gamma(\theta_s - \theta_r) \quad (9)$$

Separation Side

$$\frac{d(\sum C_{P_j} V_j \theta_s)}{dL} = -Da_s^0 \mathrm{Exp}\left\{\varepsilon_s\left(1-\frac{1}{\theta_s}\right)\right\}\delta_s f_s - \gamma(\theta_s - \theta_r) \quad (10)$$

3. Initial Conditions

$$L = 0: \quad U_C = 1, \quad V_o^0 = v_o^0/u_C^0, \quad U_I^0 = DI_r, \quad V_I^0 = DI_s$$
$$U_D = U_H = V_H = V_w = 0, \quad \theta_r = \theta_s = 1 \quad (11)$$

4. Rate Expressions

$$f_r = \frac{p_c^{1/2} - (p_D p_H/K)^{1/2}}{P_r^{1/2}(1 + 1.210 p_c^{1/2} + 1.263 p_D^{1/2})^2} \quad (12)$$

$$f_s = \frac{p_H^2 p_o}{P_s^3} \quad (13)$$

Table 2: Parameters used in calculation

$K = 1.966 \times 10^{11}\,\mathrm{Exp}(-3026/RT)$ [Pa] $P_r = P_s = 1.01325 \times 10^5$ [Pa]
$m = 1$ [−] $T_o = 723$ [K] $Da_r^0 = 100$ $DI_r = 100$ $Da_s^0 = 1 \times 10^4$
$c_p = a + bT + cT^2 + dT^3$ for all gaseous components

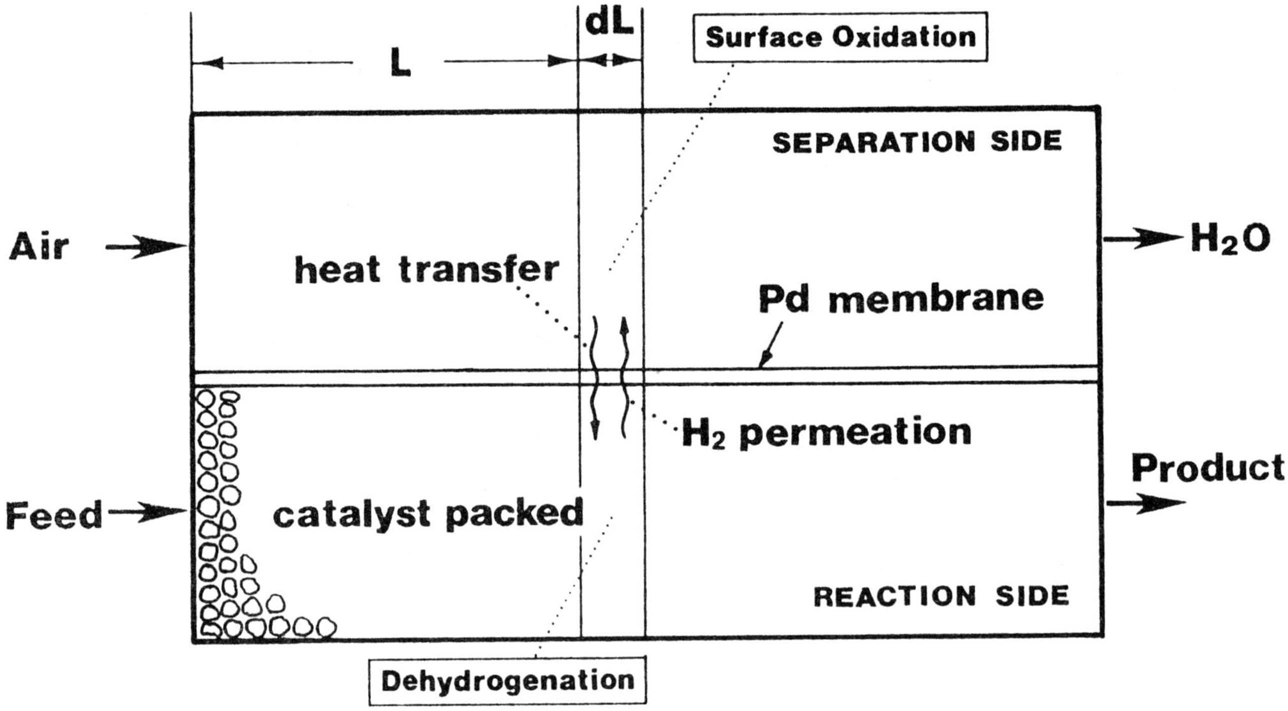

Figure 1: Model for combined dehydrogenation and oxidation by a palladium membrane reactor.

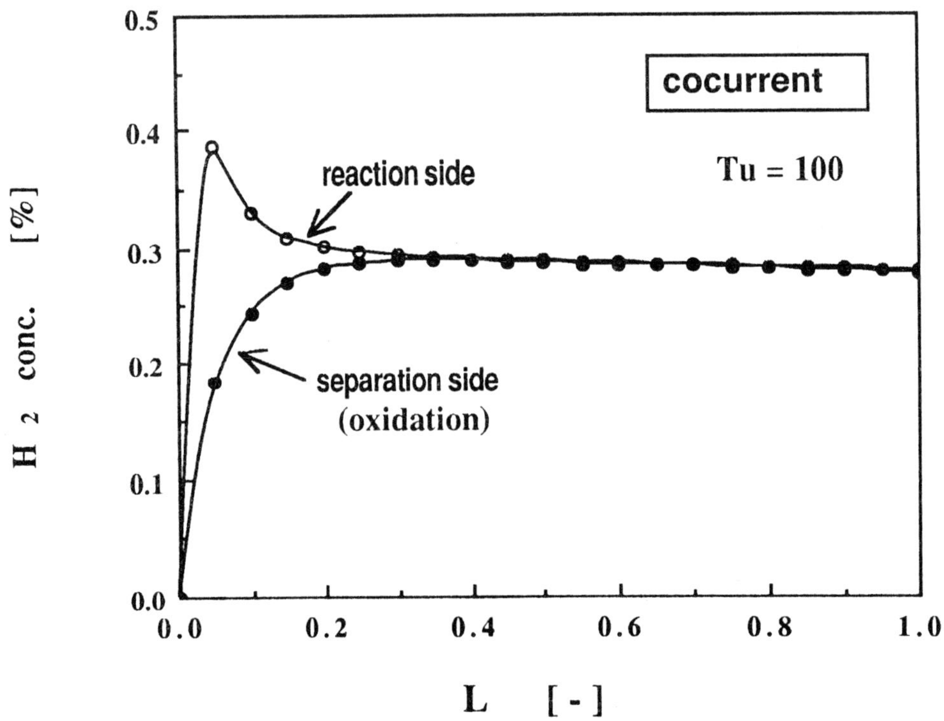

Figure 2: Hydrogen concentration profile for the co-current model.

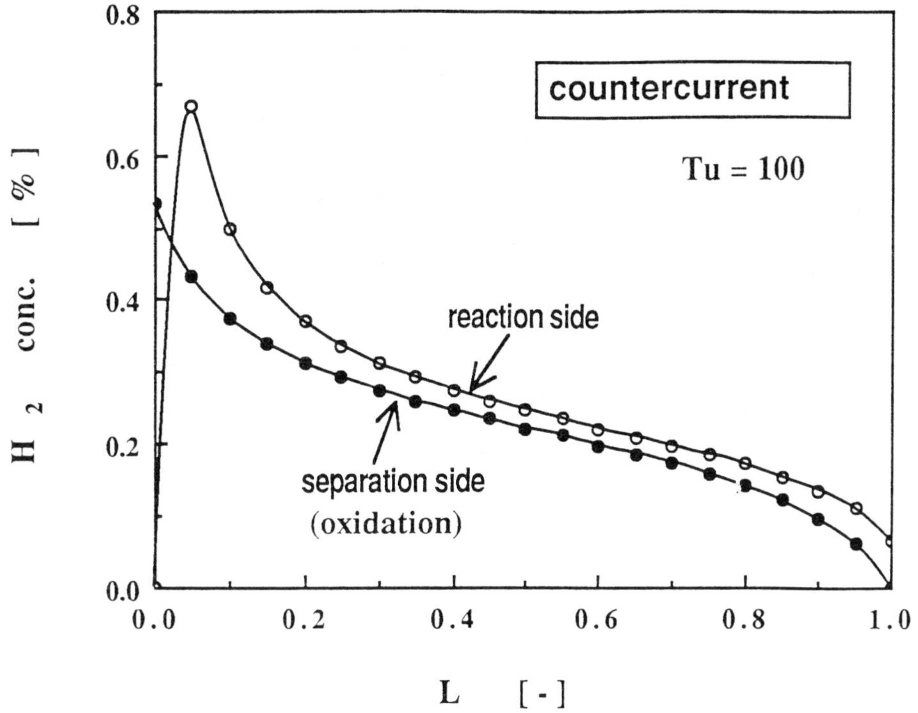

Figure 3: Hydrogen concentration profile for the counter-current case.

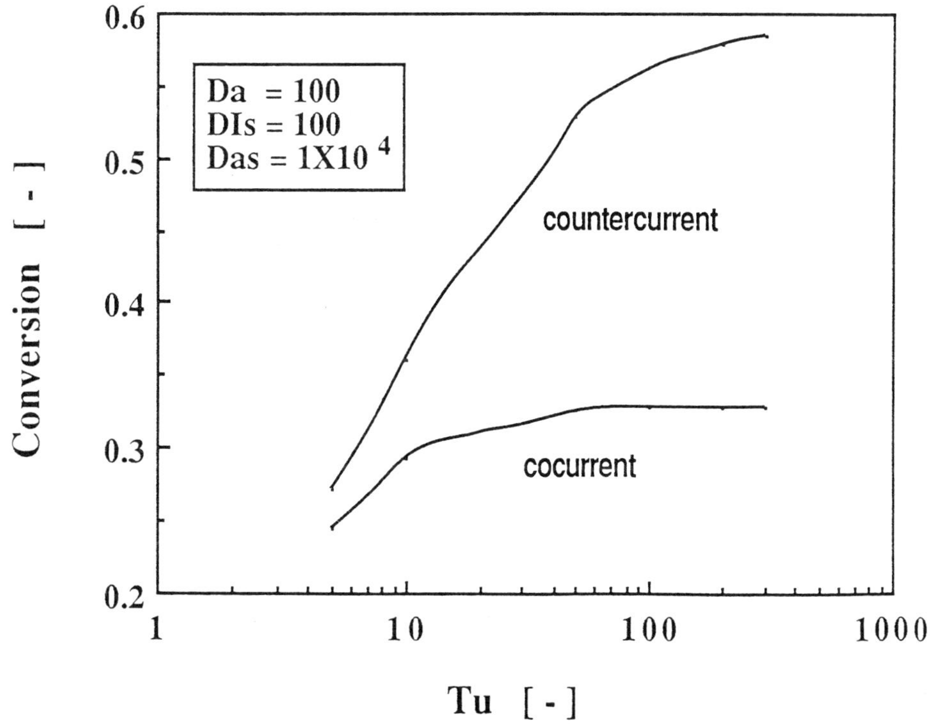

Figure 4: Comparison of performance between the co-current and counter-current models.

DEVELOPMENT OF HIGH TEMPERATURE MEMBRANES FOR MEMBRANE REACTOR: AN OVERVIEW

Shamsuddin Ilias and Rakesh Govind ■ Center-of-Excellence for Membrane Technology, University of Cincinnati
Cincinnati, OH 45221

The potential application of membranes in reactor technology and high temperature gas separation have been recognized by many investigators. Currently, a few organic polymeric membranes can withstand reaction temperatures in excess of 180 °C. However, recent developments in ceramics, inorganic membranes and dense metallic membranes have improved the opportunity to develop permselective membranes. Such membranes can be used to enhance reaction conversions for some industrially important catalytic reactions and high temperature gas separation. In this paper, the development of high temperature membranes as applied to membrane reactors is briefly reviewed.

We have identified chemical vapor deposition (CVD) and cathodic sputtering techniques as two potentially viable routes for depositing thin films on microporous supports. The application of CVD method in thin film deposition is discussed. To exploit the high selectivity of palladium and its alloys for hydrogen, we considered cathodic sputtering to deposit thin film of palladium on microporous supports. A thin palladium film was deposited on 2.5 cm porous silver disc support (0.2 mm thick and 0.45 μm pore size) substrate by cathodic Ar-ion beam sputtering technique. The ultrastructure of the film deposit indicates that the pores of the deposited film are uniform and substantially reduced in diameter. With proper choice of operating parameters of the ion beam sputter, it is possible to obtain thin film with desired pore structures. Studies are underway to characterize such films and their permselectivity.

INTRODUCTION

In recent years, membrane technology has emerged as one of the most viable unit operations in separation processes. The high technological relevance become obvious from increased application of membrane based separation in the fields of chemical processing, biotechnology & biomedical applications, environmental techniques and natural gas & oil exploitation (1, 2). The potential application of membranes in reaction engineering and high temperature gas separation are being recognized. The success of membrane in these applications will largely depend on the availability of membranes with acceptable permselectivity and thermal stability. Currently available organic membranes are not suitable for high temperature applications (> 180 °C) because of thermal degradation. On the other hand, inorganic and ceramic membranes are known for their thermal stability, but lack the permselectivity as they are highly microporous. The objective of this paper is to review the status of membrane technology and its potential application in reaction engineering. Some of the current research activity at the Center-of-Excellence to develop high temperature membrane will also be reviewed.

Correspondence concerning this article should be addressed to R. Govind.

CLASSIFICATION OF MEMBRANES

Based on the materials used for the manufacture of membranes, they can be broadly classified as organic and inorganic membranes. The first generation of membranes are the organic cellulose acetate membranes that found commercial applications in reverse osmosis (RO) and ultrafiltration (UF). However, cellulose acetate membranes have limited pH, temperature and chlorine tolerance range. Also, membrane fouling and concentration polarization are major problems affecting UF and RO operations. Cleaning of membranes to restore the flux continues to be a major task and an unsolved problem. This has lead to the development of second generation of membranes of polymeric materials such as polyamide, polysulfone, polyvinylidene fluoride and polytetrafluoroethylene (3). This class of membranes are more resistant to pH, temperature and chlorine tolerance range than the cellulose acetate membranes. They are today widely used in RO, UF and microfiltration (MF) operations. However, because of their thermal instability, these membranes are not suitable for high temperature operating environments.

Development in materials processing techniques, a new class of membranes made from inorganic compounds have emerged. These membranes are formed by deposition of inorganic solutes on microporous supports. In theory, inorganic membranes should be extremely versatile, since they can be made from various inorganic materials. Further, they should have little or none of the limitations associated with the acetate and polymeric membranes. Because of the inherent nature of inorganic molecules, this class of membranes posses a high degree of resistance to chemical and abrasion degradation, tolerate a wide range of pH and temperature. They are expected to open up a whole new area of applications. Unfortunately, the success of inorganic membranes other than UF and MF applications are severely restricted by two major factors, e.g., pore structures and permselectivity of the membranes. Currently available inorganic membranes have the pore size in the range of 40 Å to 5 μm. With this pore structure and the very nature of most of the inorganic materials, these membranes in their current form cannot be used for high temperature gas separations or for membrane reactors.

GAS SEPARATION AND MEMBRANE REACTOR

With the development of advanced material processing techniques, inorganic membranes are viewed as serious competitor to traditional gas-separation and catalytic reactor technologies. To illustrate the potential application of membrane for high temperature gas separation, consider the recovery of hydrogen from ammonia synthesis gas streams at temperatures about 450 °C. To separate hydrogen from ammonia for recycling into the reactor, today hollow-fiber polymeric membranes are used, which requires synthesis gas stream to be cooled from 450 °C to ambient temperatures. This results in significant process heat loss. With inorganic membranes, it would be possible to achieve the same separation without cooling (and thus without adding any heat exchanger) the synthesis gas stream. This will result in improving thermal efficiency (and hence the economics) of the overall process.

There are many catalytic reactions of industrial significance, where recovery or separation of hydrogen is required. Membrane technology can play an important role in these catalytic reactions. In Table 1, a partial list of catalytic reactions are given (4 - 13). Various researchers investigated the possible use of membrane reactor to perform the catalytic reaction and separation in a single process unit (14, 15). An excellent review on this subject is presented elsewhere (16). To illustrate the concept of membrane reactor, consider the catalytic dehydrogenation of cyclohexane to benzene and hydrogen. In traditional catalytic reactor, equilibrium conversion is severely limited by the presence of product hydrogen in the reaction mixture. In catalytic membrane reactor, a selective membrane is used to partition the reactor which allows the product hydrogen to permeate from the reaction side to product side as a pure component. By continuous removal of hydrogen in such a configuration, it has been demonstrated that substantial increase in conversion can be achieved.

NEED FOR HIGH TEMPERATURE MEMBRANES

As mentioned before, advances in inorganic materials processing have stimulated interest in inorganic membranes for high temperature applications in gas separation and membrane reactor. In contrast to organic membranes, inorganic membranes are able to function at higher temperatures. For example, commercially available hollow fiber membrane are limited to operation below 100 °C, while microporous membranes made of silica glass or alumina can operate temperatures as high as 800 °C. Past research on the suitability of inorganic membranes for gas separation was focused on microporous materials such as Vycor glass and alumina. Although these materials posses some permselectivity which is governed by Knudsen diffusion, the separation selectivity is very modest. This limits the applicability of these materials in its present form for high temperature application, especially in gas separation and membrane reactor where membranes must possess more selective diffusion properties.

There are few inorganic materials that have been known for selective permeation properties. For example, non-porous silica glass has high selectivity for hydrogen and helium, palladium and its alloys have high selectivity for hydrogen and silver and its alloys have high selectivity for oxygen. These materials are viewed as potential candidates for membrane materials for separations in high temperature applications.

In the past several investigators considered the

use of these non-porous inorganic membranes in membrane reactor to enhance the conversion of some industrially important equilibrium-limited reactions. The basic idea comes from thermodynamic considerations, that in a chemical reaction if one of the reaction product which slows down the reaction rate can be continuously removed, then the equilibrium of the reaction can be moved forward (equilibrium-shift), thereby enhancing the conversion. In fact, it has been demonstrated that using membranes reactors, near complete conversion of equilibrium-limited reaction can be achieved, when the rate limiting reaction product is separated and removed simultaneously (6, 7, 14).

However, the productivity of these membrane reactors are severely limited by the poor permeability of the membranes..The commercially available non-porous membranes are either thick film or thick walled tubes. Since the permeability is inversely proportional to film thickness, a thick film as membrane acts as a poor perm-separator. Thus the major challenge lies in developing a permselective thin solid film, without compromising the integrity of the film and the pores should be in the angstrom range, preferably in the range of 10 Å or less. Availability of such a membrane for high temperature application will open new area of research in membrane reactor technology and gas separation.

CHARACTERISTICS OF HIGH TEMPERATURE MEMBRANES

Membranes that could be used in high temperature applications such as gas separations and membrane reactors, must possess following characteristics:

- Thermal stability
- Nonreactive, chemically resistant
- Thermal properties, e.g. conductivity
- Permselectivity (high selectivity and high permeability)

The inorganic membranes can readily satisfy the first three requirements. However, the last requirement proposes a major limitation on the use of currently available inorganic membranes. As mentioned before, to improve the permselectivity of the inorganic membranes, one must obtain a very thin film with structural integrity and desireable thermal stability.

THIN FILM DEPOSITION TECHNOLOGY

There are various techniques by which thin solid films may be prepared. These techniques may be broadly classified as: thermal evaporation (physical vapor deposition, PVD), cathodic sputtering, and chemical vapor deposition (CVD). Reviews on various thin film deposition methods are given elsewhere (17 - 19). These reviews primarily deal with currently employed techniques in microelectronics and semiconductors. A detailed discussion of the various methods is beyond the scope of this presentation. However, we will briefly review the basic features of these processes.

In thermal evaporation, solid materials are first vaporized by heating at sufficiently high temperatures and then a thin film is deposited onto a cooler substrate by condensation of the vapor. This may be achieved directly or indirectly by a variety of physical methods, for example, resistive heating, flash evaporation, arc evaporation and laser evaporation. Thermal evaporation is usually carried under high vacuum ($\sim 10^{-5}$ tor). The technique is extensively used for deposition of thin solid films of metals such as Ag, Au, Pt and Al on various substrate.

In cathodic sputtering, material are ejected atomistically from a target by bombarding high energy positive ions (usually Ar-ions). The ejected or sputtered material is then condensed on a substrate to form a thin film. The sputtering phenomena has been known for years and exploited for deposition of films of metals and dielectrics in microelectronic fabrications. Usually in sputtering process, the high energetic particles (ions) are created in dc or rf glow discharges and then accelerated by a dc field towards the substrate. There are several methods which may be used for sputtering, e.g. glow-discharge sputtering, low pressure rf (radio frequency) & magnetic field sputtering, ion-beam sputtering and reactive sputtering.

In CVD, as the name implies, chemically reacting gases are used to synthesize thin solid films. In this process, a volatile compound of the substance that is to be deposited is vaporized and decomposed or reacted with another gas, vapor or liquid at the substrate to produce a non-volatile reaction product which deposits on the substrate as a thin solid film. Since a wide variety of chemical reactions are known for CVD applications, thus CVD technique possess the versatility and flexibility in producing thin films of pure metals and dielectrics. The various type of reactions that have been used in CVD may be broadly classified as: thermal decomposition (pyrolysis), hydrogen reduction, halide disproportionation, transfer reaction and polymerization. Most CVD processes operate in the range of few torr to above atmospheric pressure. The energy required for CVD reaction is usually provided by a thermal source, but photons or electric discharges are also used in many applications.

In Table 2, comparative summary of various methods for thin solid films is given (17). It may be seen from this table, the application of thin solid film technology is essentially in the field of protective and decorative coatings, microelectronics and semiconductors. The desired properties of thin films include good adherence, uniform thickness, impermeability and absence of crack and pinholes. Films of dielectric materials may also require thermal stability and chemical inertness.

From this brief review, it appears that the application of thin film deposition technology in fabricating thin film inorganic membranes for high temperature gas separation has never been considered as a possible route. The authors believe that some of the CVD and cathodic sputtering techniques may be used to prepare thin solid films as membranes with all desireable properties for high temperature applications.

DEVELOPMENT OF SOLID THIN FILM MEMBRANE

As mentioned earlier, for high temperature gas separation and membrane reactor, the membrane must posses high selectivity and high permeability. The high selectivity is desireable for product purity while the high permeability is essential for high product throughput from economic considerations. Currently available microporous inorganic membranes have failed to meet these requirements. However, these membranes are gaining acceptance in UF and MF applications and competing with organic membranes (14).

To develop a new class of permselective inorganic membranes, we have identified two possible routes by which inorganic membranes can be modified to meet the challenges of high temperature applications. These are: (a) deposition of thin film by CVD on microporous substrate and (b) sputtering a target material to deposit ejected atoms on a substrate as thin film.

Non-porous silica glass is known for its excellent selectivity for hydrogen. By CVD, a thin film of SiO_2 can be deposited on microporous supports, such as alumina or Vycor glass. One may use one of the reactions for the CVD as given in Table 3. It should be noted that in CVD process, the control of deposition rates and film thickness is difficult and could be a limiting factor. Besides this, other problems in the film depositions may entail dislocation formed by stresses arising in film build-up and from strains due to dissimilar thermal coefficient of expansion between the substrate and film.

A modified CVD process could be used to deposit SiO_2 into the pores of the substrate. To accomplish this, one may use a plane porous substrate to divide the reactor into two compartments and allow one of the reacting gases (say $SiCl_4$) in one compartment while the other reacting gas (say CO_2) entering the other compartment. The two reactants while diffusing through the pores in opposite direction will react to form a film of SiO_2 in the pores. In this approach, one may need to consider the thermal effects of the film in the pores (annealing and sintering) and also the possibility of pinholes due to incomplete pore blocking by the film.

However, without experiments, it is difficult to evaluate the merits of this techniques to prepare thin films for high temperature applications. Thus, future research in this direction may result in some new developments in thin film membranes.

The authors are currently experimenting with cathodic sputtering techniques to deposit thin film of Ag and Pd on microporous Ag-supports. Our choice of sputtering technique over CVD stems from three basic considerations. First, the target materials (Pd and Ag) are thermally compatible with support material (Ag) at elevated temperatures. Second, adhesion property of the film on the support is good, which is essential from the desired structural strength and film integrity. Third, the deposition rate and hence the film thickness is controllable, since in sputtering technique, the rate of deposition remains constant with time, provided current density and voltage remain constant. This is easily attained by using automatic pressure controller and a regulated power source.

The importance of film thickness can be easily shown by considering an example of permeation rate of hydrogen gas through palladium membrane (6). For isothermal, isobaric and plug flow conditions, one may approximate the permeation rate by half-power pressure law (20).

$$Q_H = k_H \left[\left(\frac{P_H^h}{P_0} \right)^{1/2} - \left(\frac{P_H^l}{P_0} \right)^{1/2} \right] \qquad (1)$$

where, P_H^h and P_H^l are the partial pressures of hydrogen on the high and low pressure sides, respectively. The permeation rate constat, k_H for hydrogen is given by:

$$k_H = \frac{A}{h} D C_0 \qquad (2)$$

where h is the membrane film thickness, A is the area normal to the flow, and D and C_0 are the diffusivity and solubility of hydrogen in palladium, respectively. For a value of $D = 9.23 \times 10^{-10}\,\mathrm{m^2/s}$ (21) and $C_0 = 1280\,\mathrm{mol/m^3}$ (22), the permeation rate constant is computed as 4.47×10^{-5} mol/s for membrane thickness of 0.2 mm. For membrane thickness of 0.001 mm, the rate constant becomes 9.94×10^{-4} mol/s, a twenty folds increase in rate constant. Thus, it can be seen from equation 1, that if the film thickness can be reduced from 0.2 mm to 0.001 mm, the productivity increases by 20 folds, provided other operating conditions remain unchanged. These reduction in thickness is achievable by various techniques that have bee described.

Thus a controlled film deposition technique such as sputtering shows a promise in this direction. In our laboratory, using ion beam sputtering technique we have deposited thin film of Pd on porous silver disc support. The silver disc was 2.5 cm in diameter (0.2 mm thick and 0.45 μm pore size). The sputtering was carried out at 10^{-5} tor pressure with Ar-ions for 60 s using a current density of 50 mA/cm^2 and an accelerating potential of 5 kV. The support was placed at 4.5 cm away from the target. The ultra structure of the support disc and deposited film are shown in Figure 1 and 2, respectively. Comparing the ultrastructure of the substrate with that of the film, it is evident that the film has a pore structure which is uniform and substantially reduced in pore size. By manipulating the operating parameters of sputtering process, it is possible to obtain a thin film with desired pore structures suitable for gas separation and membrane reactors.

Studies are underway to characterize the thin film membrane for permeability of various gases in diffusion cells, and SEM-Electro Microprobe analysis of film thickness and pore structures.

CONCLUSIONS

Membrane technology is established as one of the most viable unit operations in separation processes. With the development of new inorganic microporous membranes, which posses high resistance to chemical and abrasion degradation, tolerate a wide range of pH and temperature,they expected to open new areas of application. Currently microporous inorganic membranes are finding applications in UF and MF. Applications in areas like RO, high temperature gas separation and catalytic membrane reactors are severely limited by pore structures and poor selectivity of these membranes.

There is a need for development of new membranes suitable for high temperature gas separation and catalytic membrane reactors. It is possible to modify the pores and selectivity of inorganic membranes by thin film deposition techniques. Of these, chemical vapor deposition and sputtering techniques are identified as possible routes to deposit selective thin film membranes on microporous support. Further work is needed to realize the potentials of inorganic membranes for high temperature applications.

LITERATURE CITED

1. Hwang, S.T., and K. Kammermeyer, *Membranes in Separations,* Wiley- Interscience, New York (1975).

2. Cheryan, M., *Ultrafiltration Hand Book,* Technomic Publishing Co., Pennsylvania (1986).

3. Lloyd, D.R., *Materials Science of Synthetic Membranes,* American Chemical Society, Washington, D.C. (1985).

4. Nagamoto, H. and H. Inoue, *Bull. Chem. Soc. Japan,* 59, 3935 (1986).

5. Gryaznov, V.M., V.S. Smirnov, L.K. Ivanova, and A.P. Mishchenko, *Dokl. Akad. Nauk SSSR,* 190, 144 (1970).

6. Itoh, N., *AIChE J.,* 33, 1576 (1987).

7. Shinji, O., M. Misonou, and Y. Oneda, *Bull. Chem. Soc. Japan,* 55, 2760 (1982).

8. Sun, Y.M., and S.J. Khang, *I&EC Res.,* 27, 1136 (1988).

9. Kameyama, T., M. Dokiya, M. Fujishige and H. Yokokawa, *Ind. Eng. Chem. Fund.,* 20, 97 (1981).

10. Abe, F., European Patent Application 228,885 to NGK Insulators (1987).

11. Gul'yanova, S.G., V.M. Gryaznov and S. Kanizius, *Analiz Soverm. Zadach v Techn. Naukakh,* 172 (1973).

12. Gryaznov, V.M., V.I. Vedernikov and S.G. Gul'ynova, *Kinet. Katal.,* 27, 142 (1986).

13. Gryaznov, V.M., N.M. Ermilova, N.V. Orekhova and N.A. Makhota, *Heterog. Katal.,* 5 (Part 1), 225 (1983).

14. Mohan, K., and R. Govind, *AIChE J.,* 34, 1493 (1988).

15. Kameyama, T., M. Dokiya, M. Fujishige, H. Yokokawa, and K. Fukuda, *Int. J. Hydrogen Energy,* 8, 5 (1983).

16. Hsieh, H.P., *AIChE Symp. Ser.,* No. 261, Vol. 84, 1(1988).

17. Chopra, K.L., *Thin Film Phenomena,* McGraw-Hill Book Co., (1969).

18. Lee, S.M., *Film Deposition Techniques,* In Encyclopedia of Chemical Technology, Vol. 10, 247 (1980).

19. Jensen, K.F., *Chem. Eng. Sci.,* 42, 923 (1987).

20. Bohmholdt, G., and E. Wicke, *Z. Physik. Chem,* 56, 133 (1967).

21. Nagamoto, H., and H. Inoue, *Chem. Eng. Commun.,* 34, 315 (1985).

22. Sieverts, A., and W. danz, *Z. Physik. Chem.,* 34, 158 (1936).

Table1: Studies on catalytic reactions in membrane reactors.

Membrane Material	Permeable species	Catalytic reaction	Reference
Pd	H	Hydrogenation of butadiene	4
Pd	H	Dehydrogenation of butene to butadiene	5
Pd	H	Dehydrogenation of cyclohexane	6
Glass	H	Dehydrogenation of cyclohexane	7, 8
Glass	H	Decomposition of H_2S to H_2S and S	9
Al_2O_3	H	Decomposition of H_2S to H_2S and S	10
Pd-Ag	H	Hydrogenation of acetylene	11
Pd-Rh	H	Dehydrogenation of n-heptane	12
Pd-Sb	H	Hydrogenation of butenes	13
Al_2O_3/SiO_2	H	Decomposition of H_2S to H_2S and S	10

Table 2: Some characteristics of various thin film depositiuon techniques (17).

Technique	Materials	Deposition rate*	Rate control	Vacuum (μ)	Adhesion	Remarks
Evaporation						
Resistive heating	Most	L-M	Yes	<0.1	Fair	Simple, support alloying and deposition in certain cases.
Flash	Alloys	M-H	No	<0.1	Good	Nearly Stoichiometric film, gas burst, spattering.
Arc	Conducting	H	No	<1	Fair	Crude, very high rate, spattering.
Laser	Most	H	No	<0.1	Fair	Very high rate, spattering, promising method.
Sputtering						
Glow discharge	Conducting	L-M	Yes	20-60	Good	Simple, widely used.
RF	Most	L-H	Poor	~ 1	Good	Useful, plasma effects complicated.
Ion beam	Most	L-H	Poor	<0.1	Good	Sophisticated, welldefined conditions, low pressure, substrate out of plasma.
Reactive	Oxides, hydrides, Carbides, nitrides sulphides	L-M	No	1-60	Good	Powerful, many variables, difficult to control.
CVD						
Pyrolysis	Most	H	No	~1000	Fair	Production method, high temperature system, contamination problem.
Hydrogen reduction	Most	H	No	~1000	Fair	Production method, high temperature system, contamination problem.
Halide disproportionation	Products of some halides	H	No	~1000	Fair	High temperature, contamination problem.
Transfer reaction	Depends on reaction	H	No	~1000	Fair	High temperature, contamination problem.
Polymerization	Polymers	H	No		Fair	Cheap, good films, thermally unstable.

* L, M, H stands for low (<0.1), medium (0.1 to 20) and high (>20) rates of deposition (Angstrom/sec), respectively.

Table 3: Typical SiO_2 film deposition reactions for atmospheric CVD reactors.

Carrier gas	Film deposit	Deposition temperature range (°C)	CVD reaction
H_2	SiO_2	600-900	$SiH_4 + 2CO_2 \rightarrow SiO_2 + 2H_2 + 2CO$
H_2	SiO_2	500-900	$SiH_4 + 4CO_2 \rightarrow SiO_2 + 2H_2O + 4CO$
H_2	SiO_2	800-1000	$SiCl_4 + CO_2 \rightarrow SiO_2 + CCl_4$
H_2	SiO_2	200-500	$SiH_4 + O_2 \rightarrow SiO_2 + 2H_2$.

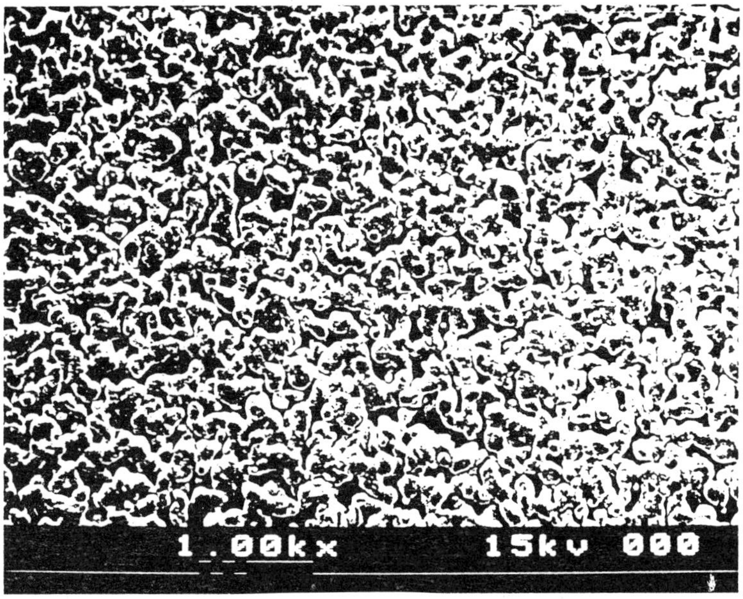

Figure 1: Ultrastructure of microporous silver substrate (membrane support), 1K magnification.

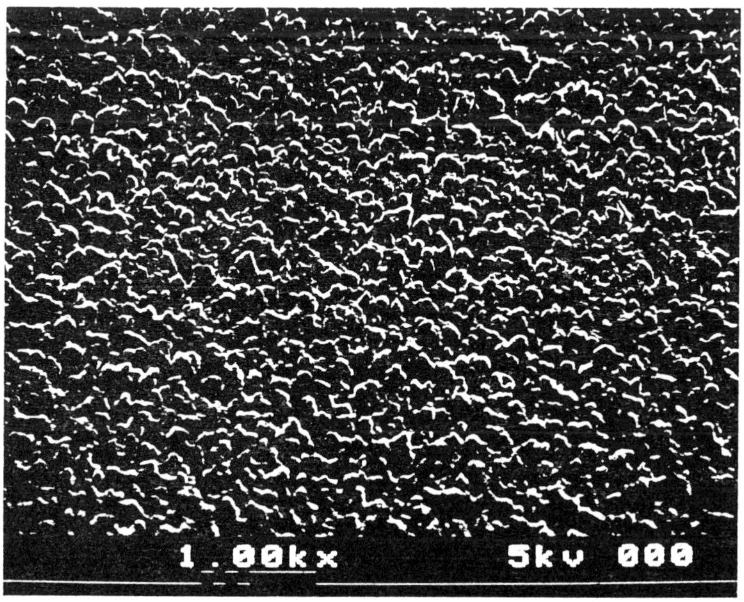

Figure 2: Ultrastructure of palladium thin film on silver substrate deposited by cathodic sputtering, 1K magnification.

THE CATALYTICALLY IMPREGNATED CERAMIC TUBE: AN ALTERNATIVE MULTIPHASE REACTOR

Michael P. Harold, Philippe Cini
and Bruce Patenaude ■ Department of Chemical Engineering, University of Massachusetts, Amherst, MA 01003
Venkataraman ■ Norton Company, New Bond Street, Worcester, MA 01606-2698

This study investigates the use of hollow ceramic tubes as multiphase catalysts. The catalytically impregnated tube serves as a gas-liquid contacting device. The gas flows through the tube core and the liquid flows on the shell side and fills the pores in the tube wall, within which the catalytic reaction occurs. This design gives reduced transport limitations, better exothermic heat control, and well defined segregated two-phase flow.

In this study a critical performance comparison is made between the catalytic tube and a string of suspended spherical catalytic pellets, considered the prototype trickle-bed reactor. Heterogeneous models are developed for (i) the string of pellets contacted by cocurrent downflow of the gas and a liquid film, and for (ii) the single tube with gas flow through the tube core and liquid flow both cocurrent annular and well mixed on the shell side. Simulations of a volatile reactant limited liquid phase catalytic reaction reveals that for an identical catalyst volume the catalytic tube efficiency exceeds that of the pellet string by as much as an order of magnitude. The key advantage is that at high liquid flow rates the inside wall of the tube remains directly exposed to the gas whereas the pellets are completely wetted, thus partially eliminating the liquid film mass transfer resistance. The impacts of several variables are assessed, including the degree of liquid phase saturation by the volatile reactant, the catalytic activity, and the catalyst dimensions.

INTRODUCTION

Many commercially important chemicals are products from the catalytic reaction between volatile and nonvolatile compounds. Two major difficulties are encountered in carrying out these multiphase reactions; a limited solubility of the volatile reactant in the liquid phase and both intraparticle and interparticle mass transport limitations. For example, noble metal catalyzed olefin hydrogenations are controlled by mass transport because the catalysts are often quite active.

Most multiphase catalytic reaction systems are carried out in the slurry, ebullated bed, fixed bed bubble, or trickle-bed reactor (Tarhan, 1983). The cocurrent downflow trickle-bed design is the usual choice for systems which do not require frequent catalyst regeneration. Although several gas-liquid flow patterns are encountered with this design (Sato et al., 1973; Ng, 1986), most common in practice is trickling flow in which the liquid flows as rivulets and films over the catalyst while the gas flows through the remaining voids. For sufficiently low liquid flow rates some of the pellets are only partially wetted by the liquid. The flowing liquid film supplies the nonvolatileand dissolved volatile reactants to the catalyst pores. However, it also deters the supply of the volatile reactant from the gas phase. In general, the supply of the volatile reactant to the liquid filled pores is much slower than it would be directly from the gas at the nonwetted surface. For this reason, partially wetted catalysts can perform better than fully wetted ones for volatile reactant limited reactions (Herskowitz et al., 1979; Yentekakis and Vayenas, 1987; Harold and Ng, 1987). This is called effectiveness enhancement (Harold and Ng, 1987). Effectiveness enhancement explains the experimental observation in trickle-bed studies of a declining rate with increasing liquid flowrate (Satterfield et al., 1969; Herskowitz et al., Germain et al., 1978); i.e., as the flowrate increases more of the external catalyst surface is wetted and the supply of the volatile reactant declines.

M. Harold and P. Cini are with Dept. of Chemical Engineering at the Univ. of Massachusetts.
B. Patenaude is currently at the Dept. of Chemical Engineering at the Univ. of Houston.
K. Venkataraman is with the Norton Company in Worcester, Massachusetts.

Clearly, for mass transport controlled, volatile reactant limited reactions the gas-liquid contacting strategy and the catalyst size and shape are critical variables in the reactor design. A key challenge is whether one can exploit these variables in a controlled fashion in order to capitalize on the more rapid volatile reactant transport from the bulk gas than from the bulk liquid. The main goal is to develop a novel reactor that performs better than the conventional trickle-bed.

A few pioneering studies have confronted this challenge. In the first of the approaches de Vos et al. (1982) converted a standard cross-flow heat exchanger into a catalyst to carry out the liquid phase hydrogenation of nitrobenzoic acid and a complex mixture of unsaturates. A model was developed for this configuration by de Vos and Hamrin (1982). Akyurtlu et al. (1985, 1988) developed a similar model for the cylindrical catalyst geometry with cocurrent flow of gas and liquid through the tube- and shell-side, respectively. They hinted that this design might be more attractive than the cross-flow arrangement. Unfortunately, the modelling studies did not compare the performance of the segregated flow designs with the trickle-bed reactor. The second group of studies involved the use of a porous catalytic ceramic monolith (Hatziantoniou and Andersson, 1982, 1984; Hatziantoniou et al., 1986a). The segmented flow was shown to possess better mass transport features than for the continuous liquid flow mode, and lower pressure drops compared to the trickle-bed (Satterfield and Ozel, 1977). The third study involved involved the use of a polymeric (PTFE) membrane which was coated on one side with a catalytic powder (Hatziantoniou et al. 1986b).

Our proposal is to develop the catalytically impregnated ceramic tube as a new type of multiphase reactor. We have selected the hollow tube geometry because ceramic (α - Al_2O_3) tubes are used successfuly as supports in some gas phase catalytic applications and as microporous filters. Thus, the technology to make the tubes is in place. The key step is to convert these ceramic tubes into active catalysts.

The operational strategy is to pass the gas through the tube core while immersing the tube in the flowing liquid stream, which surrounds the shell side and fills the tube pores. As mentioned earlier, the catalyst pellets within the trickle-bed are partially wetted at sufficiently low liquid flow rates and completely wetted if the liquid flow rate is high enough. On the other hand, the catalytic tube is only wetted on one side for the entire range of liquid flow rates. Thus, the enhancing effect of the supply of volatile reactant directly into the pores from the gas is achieved over a much larger range of conditions. Moreover, compared to the trickle-bed design, the tubular design gives well-defined, segregated gas-liquid contacting, and potentially better heat control for exothermic reactions.

The first step of this research is to show via detailed modelling that the catalytic tube design is an attractive alternative to the trickle-bed reactor. A rational comparison is essential in order to locate the conditions for which performance improvements can be obtained and to calculate the magnitude of the improvements. This paper describes our approach and results for this step. The second and more critical step is to develop techniques for preparing active and durable catalytic tubes for use in a multiphase reaction system. For example, although α-Al_2O_3 is extremely strong, its low surface area and lack of hydroxyls in its lattice makes it a rather poor catalyst support for most reactions. Thus, the support needs to be modified, such as coating the tube with a high surface area γ-Al_2O_3 layer. This second step is addressed in future studies.

MODEL DEVELOPMENT AND PREDICTIONS

Our approach in this paper is to compare and contrast the performance of a single catalytic tube exposed to cocurrent, segregated gas-liquid flow or segregated gas - well mixed liquid flow, with a prototype trickle-bed reactor, a vertical suspended string of spherical catalyst pellets exposed to cocurrent gas-liquid film flow.

The reaction of interest is given by

$$A_{(g)} + \nu B_{(l)} \rightarrow \text{Product(s)}_{(l)}$$

where g(l) denotes gas(liquid). The reaction is assumed to have negligible heat effects. The intrinsic rate(r) is assumed first-order in the volatile reactant (A) and zeroth-order in the nonvolatile reactant (B):

$$r = k' c_{Bf} c'_A H'(c'_B) = k c'_A H'(c'_B) \qquad (1)$$

where c'_A (c'_B) is the intraparticle concentration of A(B), k' is the bimolecular rate constant, c_{Bf} is the bulk liquid phase concentration of B, with $k = k'c_{Bf}$ the pseudo first order rate constant, and H' is the Heaviside function. These kinetics are representative of a large class of liquid phase oxidations and hydrogenations.

We now describe the essential ingredients of the three reactor models. The developed governing equations appear in Appendices A, B, and C which are in the Supplementary material section. The reader is referred to a future publication for a more detailed presentation (Cini and Harold, 1989).

String of Pellets Reactor

A schematic of the string of pellets is depicted in Figure 1.a. The N spherical pellets of radius R_p are suspended in a tube shell of radius R_s. The total length of the pellet string is L (=2NR_p). The liquid is injected over the top pellet and gas flows cocurrently in the annular region. The nonvolatile reactant (B) is supplied from the liquid film and the volatile reactant (A) is supplied from the gas, which contains A and an inert carrier (I). The liquid also contains B, reaction product(s), and a solvent. The liquid enters devoid of A, partially or fully saturated with A. The pellets are partially or fully wetted by the liquid depending on the magnitude of the volumetric liquid flowrate (q_f). The critical volumetric flowrate, q_f^o, below (above) which partial (complete) wetting occurs can be estimated by the dry patch model of Hartley and Murgatroyd (1964) applied to flow over a single sphere:

$$q_f^o = \frac{2\pi R_p \Gamma_o}{P_f} = w\, 2\pi R_p \left(\frac{\mu_f \sigma^3}{\rho_f^4 g} \right)^{1/5} \qquad (2)$$

Γ_o is the critical mass flowrate per unit width, w is a constant of order 0.1 (Norman and McIntyre, 1960), μ_f and ρ_f are the liquid viscosity and density, σ is the interfacial tension, and g is the gravitational acceleration. We assume that the degree of wetting, the wetting efficiency (E_w), is given by q_f/q_f^o.

Our approach is to treat the string of spherical pellets as a catalytic wall of thickness δ contacted by the liquid (Figure 1.b). As long as the appropriate characteristic diffusion length for a sphere is used, this approach is reasonable and is especially convenient from a modelling standpoint.

The string of pellets model consists of a species A mole balance and a momentum balance within the gas filled annular region, balances for A and B in the liquid film, and balances for A and B within the catalyst. These are provided in dimensionless form in Appendix A, Eqns. (A.1 - A.16).

The equations which describe the external mass transport, intraparticle diffusion, and reaction for the partially or fully wetted slab catalyst are from the model of Harold and Ng (1987), which is now briefly summarized. The partially wetted catalyst is subjected to a nonuniform environment along its surface. Thus, formally the intraparticle diffusion and reaction is three-dimensional in nature; i.e. there are gradients in the directions parallel (x) and normal (y) to the wetted surface and in the direction of the liquid film flow (z). The latter are expected to be negligible over a single pellet. The Harold and Ng model is an approximate one-dimensional formulation which accounts for nonvolatile reactant (B) depletion.

As shown in Figure 1.b, the slab shaped pellet is divided into two regions; a part below the wetted surface ($0 < x < E_wS/2$) and a part below the nonwetted surface ($E_wS/2 < x < S/2$), where S is the total length of the exposed pellet surface. Depending on the conditions, species B may not deplete, may deplete within the nonwetted or wetted parts, or may deplete completely within the pellet. Eqns. (A.6 - A.11) apply to the case in which B depletes within the nonwetted part at a point $s^* = x^*/(S/2)$. The intraparticle diffusion process in the x direction (or equivalently the s direction, where $s = 2x/S$) provides the communication between the wetted zone ($0 < s < E_w$), the nonwetted zone in which B is present ($E_w < s < s^*$), and the nonwetted zone in which B is depleted ($s^* < s < 1$). The diffusion process in the direction normal to the surface within a given region and the external transport for the same region is lumped into the same term consisting of the product of an overall Biot number and concentration difference. For example, species A intraparticle diffusion in the normal direction within the wetted part and external transport from the liquid film is accounted for by the second term in Eqn. (A.6) where the overall liquid-to-solid Biot number, Bi_{lsA}, is given by Eqn. (A.14). By applying the appropriate boundary conditions (A.26 - A.31) analytic expressions are obtained for the intraparticle A and B concentration profiles in each region. The depletion point (s^*) appearing in the solution is calculated by solving a single implicit function that results by demanding the the species B concentration vanish at $s = s^*$ (Eqn. (A.30.a)). One change has been made in the Harold and Ng model for application to the problem at hand. The original single pellet model lumps the gas-to-liquid mass transfer resistance with the liquid-to-solid resistance. The current model cannot because of its heterogeneous formulation.

The momentum balance (A.1) contains convective terms, pressure gradient terms, and a viscous shear term. Plug flow behavior is assumed in formulating the momentum balance and the gas phase mole balances for A and I (Eqns. (A.2) and (A.3)). Eqn. (A.2) is continuity of species I, which is assumed insoluble in the liquid. Eqn. (A.3) contains the analogous convective term for A and three more terms which respectively account for transport of A into the liquid film, into the nonwetted part of the pellet in which B is present, and into the nonwetted part of the pellet in which B is depleted. Similarly, the liquid phase balances for A and B (Eqns. (A.4) and (A.5)) contain convective terms and exchange terms. Note that the balances are somewhat different if B does not deplete or depletes within the wetted part.

The external mass transport coefficients k_{lsA}^o, k_{lsB}^o, and k_{gl} can be estimated from expressions developed by Norris and Streid (1940) for liquid-to-solid transport and by Lynn et al. (1952) for gas-to-liquid transport. These are summarized in Eqns. (A.17 - A.23). Inspection of Eqns. (A.17) - (A.22) reveals that $k_{gl} \alpha q_f^{1/3}$ and $k_{lsB}^o \alpha q_f^{1/9}$. The latter dependence agrees with the analytical result for liquid-to-solid mass transport during film flow down a soluble inclined wall (Bird et al., 1960). Note that the gas-to-solid coefficient for A, k_{gsA}^o, is typically much larger in magnitude than the other coefficients (Morita and Smith, 1978; Herskowitz et al., 1979).

The system of first-order differential equations (in Z direction) are solved subject to the inlet conditions (A.24) and (A. 25). The integration is carried out by a semi-implicit scheme (subroutine STIFF provided by Villadsen and Michelson, 1978). The integrals which appear are given by analytic expressions. During each function evaluation at any location in the reactor the single pellet model is used to calculate the intraparticle profiles and the depletion point, s^* (if necessary). The dimensionless A and B film concentrations appearing in the single pellet model, u_{Af} and u_{Bf}, are related to the variables of integration by $u_{Af}(Z) = c_{Af}(z)/(c_{AG}(z)/H)$ and $u_{Bf} = c_{Bf}(z)/c_{AG}(z)/H$.

Some quantities of interest include the fractional conversions of the volatile reactant A, X_A, nonvolatile reactant B, X_B, and the overall effectiveness, or efficiency, of the pellet string, E. The conversions X_A and X_B have the standard definition and are given by Eqns. (A.32) and (A.33), respectively. The efficiency is defined as the overall A consumption rate normalized by the consumption rate with no transport limitations; i.e., the liquid phase concentration of A is everywhere equal to its equilibrium solubility of A based on its inlet gas phase concentration ($c_{AG}(z = 0)/H$), and the concentration of B is everywhere equal to its inlet liquid phase concentration ($c_{Bf}(z = 0)$).

Model Simulations. Considering the large number of parameters it is impractical to do a complete parametric sensitivity study. Instead, three parameters are varied to illustrate several key trends; the liquid flowrate (q_f), catalytic activity (k'), and the inlet dissolved volatile reactant concentration ($c_{Af}(z = 0)$). Table 1 provides the values of the dimensionless parameters for the analysis. In most of the plots the pellet string efficiency (E) is calculated as a function of the liquid flow rate normalized by the critical liquid flowrate which divides the partial and full wetting regimes, $Q = q_f/q_f^o$. The parameter P_f, which is proportional to q_f, is adjusted appropriately as Q is varied. Also, the external mass transport coefficients are calculated using Eqns. (A.17) - (A.23). The catalytic activity affects only the value of the Thiele modulus based on the inlet concentration of the nonvolatile reactant, ϕ_o. The species A inlet liquid concentration, $c_{Af}(z = 0)$, is varied by assigning $C_A{}^o$ and $C_f{}^o$ the four pairs of values shown in Table 1.

Figure 2 shows how the efficiency (E) depends on the liquid flow rate (Q) for several values of ϕ_o (0.1, 10, 50, 100, 200). For all curves $C_A{}^o = 0.5$ (half-saturated liquid feed) and $C_f{}^o = 50$ (large excess of B in feed liquid). The curves share some similar features. First, for Q ---> 0 the pellets are practically nonwetted and the nonvolatile reactant, which is supplied only from the liquid film, is the limiting reactant. That is, B depletes within the pellet for a sufficiently small Q and E(Q ---> 0) ---> 0 even though the inlet concentration of B exceeds the species A solubility by a factor of 50. Second, the efficiency is a rather insensitive function of Q for Q > 1 due to the weak dependencies of the mass transport coefficients $k_{lsA}{}^o$, $K_{lsB}{}^o$, and k_{gl} on q_f. Third, each curve exhibits a local maximum efficiency.

The efficiency maximum is a result of different mechanisms depending on the catalytic activity. For the inactive pellets ($\phi = 0.1$) the efficiency maximum occurs for flow rates approaching zero. The efficiency decreases sharply from its maximum and levels off at 0.5 for high flow rates. This demonstrates that at low liquid flow rates the volatile reactant is able to closely approach its equilibrium solubility limit throughout all the pellets. This is especially true for the partially wetted catalyst because of the more efficient supply of A through the nonwetted surface. As Q increases the pellets become completely wetted and the concentration of A approaches its inlet unsaturated liquid phase value of 0.5.

If the catalytic activity is increased the efficiency decreases for all flow rates. Although an increase in ϕ_o from 0.1 to 1 has negligible effect, an increase to 10 decreases E in the fully wetted regime from about 0.5 to 0.04. A further increase ϕ_o to 100 reduces E to about 0.004. The fact that E is inversely proportional to ϕ_o indicates intraparticle diffusional limitations. The flow rate giving the maximum efficiency within the partial wetting regime increases as ϕ_o increases. For the active catalyst the local efficiency maximum for Q < 1 is a result of effectiveness enhancement; i.e., the more rapid supply of volatile reactant into the pores through the nonwetted surface than the wetted surface results in an increase in the efficiency.

The degree of inlet liquid saturation by the volatile reactant can have a significant impact on the pellet string performance. Most previous trickle-bed models have not considered this key variable. Figure 3 shows a plot of E versus Q for four different inlet liquid concentrations of species A (see Table 1). For all curves $\phi_o = 100$. The complete saturation ($C_A{}^o = 1$ and half-saturation ($C_A{}^o = 0.5$) curves have features similar to the diffusion controlled curves of Figure 2. The effectiveness enhancement is not as evident for the $C_A{}^o = 1$ case because the liquid does not pose as much a resistance to the supply of A into the pellet. As expected, E decreases for all flowrate levels as the degree of saturation decreases. A sufficiently low degree of saturation significantly reduces the efficiency. For the $C_A{}^o = 0.01$ case the efficiency decreases over an order of magnitude from its maximum value at very small Q. In this case the liquid acts as a barrier to volatile reactant supply; the faster the flow the more detrimental the impact.

Simulation of Experimental Data. It is instructive to test the predictive capabilities of the string of pellets model. Satterfield et al. (1969) carried out the hydrogenation of α-methylstyrene (AMS) on a suspended string of fourteen spherical palladium on alumina supported catalytic pellets. Pure hydrogen at about 1 atm flowed cocurrently with hydrogen saturated AMS over the string of 0.825 cm diameter pellets. The overall rate per unit volume was measured as a function of the inlet volumetric liquid flowrate (q_f) for several temperatures between 20 and 50 °C. The 50 °C data (solid squares in Figure 4) demonstrates nonmonotonic trends. As q_f is increased from zero the rate increases to a maximum at about 0.13 cm^3/sec. Between this flowrate and about 0.36 cm^3/sec the rate declines. Further increases in q_f to about 0.65 cm^3/sec results in a gradual rate increase. The investigators contended that the rate increase with decreasing flow rate was essentially a transient, and not a steady-state phenomenon. The rate increase for $q_f > 0.36$ cm^3/sec was attributed to external transport effects.

The trends observed are entirely consistent with the features predicted by the string of pellets model. For a sufficiently low liquid flowrate the pellets are only partially wetted by the flowing liquid. Within the partially wetted regime the rate attains a maximum because of the effectiveness enhancement phenomenon. In the fully wetted regime the rate is a weak increasing function of the flowrate due to external mass transport effects. Thus the existence of two local extremum. Our contention of partial wetting is supported by observations made by the investigators. During some holdup experiments using room temperature water, incomplete wetting of 0.825 cm diameter nonporous glass spheres was observed for flowrates below 0.5 cm^3/sec. This observation can be approximately translated to the reaction experiments as follows. If the 0.5 cm^3/sec flowrate is considered the critical flow rate using room temperature water, $q^o_{f,water,20}$, and any difference between the wetting behavior on the glass spheres compared with the porous alumina supported pellets is ignored, then the critical flow rate using AMS at 50°C, $q^o_{f,AMS,50}$ is estimated as follows, using the Hartley-Murgatroyd criterion (eqn. (2)):

$$q^o_{f,AMS,50} = q^o_{f,\ wat.,50}\left(\frac{\mu_f \sigma^3}{\rho_f}\right)^{1/5}_{AMS,50}\left(\frac{\mu_f \sigma^3}{\rho_f}\right)^{-1/5}_{wat.,50} \qquad (3.a)$$

$$q^o_{f,\ wat.,\ 50} = q^o_{f,wat.,\ 20}\left(\frac{\mu_{f,\ wat.,\ 50}}{\mu_{f,wat/,\ 20}}\right)^{1/5} \qquad (3.b)$$

The 50 °C AMS properties are approximately $\mu_f = 0.135$ cP (as predicted by the equation of Thomas, 1946), $\rho_f = 0.908$ g/cm^3, and $\sigma = 30$ dyne/cm. The 50 °C water properties are $\mu_f = 0.547$ cP, $\rho_f = 1$ g/cm^3, and $\sigma = 70$ dyne/cm. Substitution of these values into Eqns. (3.a) and (3.b) gives $q^o_{f,\ AMS,\ 50} = 0.217$ cm^3/sec. Substitution of this value and the 50 °C AMS properties into Eqn. (2) gives w = 0.149. This value is rather close to 0.1, the correct order of magnitude.

Inspection of the experimental data (Figure 4) reveals a reasonably good agreement between the 0.217 cm^3/sec and the location of the local rate minimum, which qualitatively at least could be considered the point at which the pellets become fully wetted.

The real test of the pellet string model is to use this estimate of the partial wetting regime and the provided data to simulate the entire 50 °C experiment. Table 2 provides the kinetic, transport, and physical property values that we used for the simulation. Most of the numbers were provided by Satterfield et al. (1969). The values we estimated are marked with an asterisk. The catalysts used were of the "egg shell" variety; a 0.1625 cm radius core was inert ($R_{p,c}$ = 0.1625 cm) and the outer shell of 0.25 cm thickness was impregnated with Pd. The sphere-equivalent active layer thickness is then given by $\delta = aR_p/3$, where $a = 1 - (R_{p,c}/R_p)^3$ is the fractional volume of active material. Note that the pellet string efficiency E based on a uniformly impregnated catalyst is modified according to Eqn. (A.35). The coefficients k_{gl}, $k_{lsA}{}^o$, and $k_{lsB}{}^o$ are calculated using Eqns. (A.17) - (A.23). The hydrogen gas-to-solid coefficient, $k_{gsA}{}^o$, is assumed to be some factor ($\kappa > 1$) greater than k_{gl}.

The first calculation is to check the predicted single pellet effectiveness with no external transport limitations. Satterfield et al. measured a rate in a spinning basket reactor of 3.42 x 10^{-7} g mole/(sec) (cc pellet) for the full size pellets. Using the appropriate values in Table 2, the single pellet model predicts a value which is only 2.2% higher. The second test of the model is to simulate the rate versus liquid flowrate data. The solid line in Figure 4 is the prediction of the pellet string model. The ratio $\kappa = k^o_{gsA}/k_{gl}$ is set equal to 5. The model predicts all the essential nonmonotonic trends and the correct magnitude of the rate. Of course, the values of w, the factor in the Hartley-Murgatroyd criterion, and κ could be adjusted to give a better data fit. However, the key result to note is that the model capably predicts the transitions from the AMS depletion to non-depletion regimes and the partial to complete wetting regimes. These results indicate that the catalytic slab model is a reliable simulator of a string of spherical pellets.

COCURRENT FLOW CATALYTIC TUBE REACTOR

Figure 5 shows a schematic of the catalytic tube of inner radius R_1, outer radius R_2, and length L positioned concentrically within an outer shell of radius R_s. The gas and liquid flow cocurrently in the tube core and in the annular region, respectively. It is assumed that the catalyst pores are completely filled with the liquid, and that the gas-liquid interface is stabilized at the pore mouths located at the inner tube wall. The gas contains an inert I and the volatile reactant A; the liquid contains A, the nonvolatile reactant B, the products, and possibly a solvent.

The model consists of a momentum balance and species I and A balances in the gas core (Eqns. (B.1) - (B.3)), reaction-diffusion balances for species A and B within the tube wall (B.4) and (B.5)), and species A and B balances in the liquid filled annulus ((B.6) - (B.7)). The gas phase momentum balance (B.1) is identical in form to the one for the pellet string model (eqn. (A.1)) except that it applies to tubular rather than annular flow. The species A balance (B.3) includes a plug flow type convective term and a term which accounts for absorption into the tube wall, assuming a negligible mass transport resistance. The intratube balances (B.4) and (B.5) include terms for effective molecular diffusion in the radial direction and reaction. Axial directed intraparticle transport is expected to be negligible compared with the bulk phase convection. Note that the case exhibited in Appendix B is for no depletion of B within the tube. For a sufficiently active catalyst the supply rate of B from the liquid may be insufficient, causing B to deplete at some intermediate location within the wall ($r = r^*$). In the limit of an extremely active catalyst, B may deplete completely. In such a case the reaction occurs instantaneously at the $r = R_2$ surface. Finally, the liquid phase balances (B.6) and (B.7) contain axial convection terms assuming laminar flow that is uncoupled from the mass transport, and radial and axial dispersion terms. In our analysis we neglect the axial dispersion; this is expected to be a good assumption for non-vanishing flow rates and a tube with length L sufficiently larger than the annular gap (R_s - R_2). The appropriate boundary conditions (B.8) - (B.13) complete the model.

A similar model was recently developed by Akyurtlu et al. (1988). However, there are two key differences. Our model includes a variation in the gas phase composition, velocity, and pressure; their model assumes an isobaric gas containing pure volatile reactant along the entire tube length. Our model correctly accounts for the variation of the pseudo first-order rate constant with the nonvolatile reactant B bulk concentration; i.e., $k(z) = k'c_{Bf}(r = R_2, z)$; their model assumes that the pseudo first-order rate constant is independent of the species B concentration.

The model consists of two ordinary differential equations in the gas core and two partial differential equations in the liquid annulus. These are coupled via the reaction-diffusion balances and boundary conditions. The model is solved by discretizing the liquid phase balances in the r direction by orthogonal collocation (Villadsen and Michelsen, 1978). The system consists of 2n + 4 ordinary differential equations, where n is the number of interior collocation points, two nonlinear algebraic equations involving the flux boundary conditions at $r = R_2$ (Eqns. (B.10) and (B.11)), and another function of the depletion point (ξ^*; if depletion occurs). The equations are solved by a semi-implicit integration scheme modified to accommodate the nonlinear boundary conditions. At each function evaluation the depletion point (if necessary) and fluxes at R_1 and R_2 are calculated.

The performance indicators of interest, the species B and A fractional conversions and the catalytic tube efficiency are given by Eqns. (B.14) - (B.16), respectively. The conversion expressions are derived from the standard definitions and the efficiency is defined in the same way as in the pellet string model.

Model Predictions. It is instructive to compare the cocurrent flow catalytic tube performance to the string of pellets by calculating the tube efficiency as a function of the liquid flowrate for different catalytic activities (ϕ_o) and the inlet liquid concentration of the volatile reactant ($C_A{}^o$). Table 2 lists the parameter values for the simulations. The flowrate (q_f) is varied by adjusting the liquid phase Peclet numbers Pe_{fA}' and Pe_{fB}', both of which are proportional to q_f.

Figure 6 shows how the efficiency (E) varies with the species A liquid phase Peclet number (Pe_{fA}') for three ϕ_o values (0.1, 5, 10) and a saturated liquid ($C_A{}^o = 1$; solid lines). Each curve is monotonic increasing. For a fixed flowrate (Pe_{fA}') the efficiency is a decreasing function of ϕ_o. For the lowest activity ($\phi_o = 0.1$) E is close to unity for all nonvanishing flowrates, demonstrating a virtual absence of transport limitations. Although it is difficult to see on the logarithmic scale of Figure 6, as the flowrate approaches zero the efficiency decreases to zero as well. An increase in ϕ_o from 0.1 to 5 leads to a sharp efficiency decrease, indicating the onset of transport limitations. In fact, the tube is partially devoid of B for the $\phi_o = 5$ and 10 cases ($\xi^*(Z = 0) = 2r^*/(R_2 - R_1) = 3.67$ (3.77) for $\phi_o = 5$ (10); with $Pe_{fA}' = 100$). This is a notable disadvantage of the tube design compared to the string of pellets; i.e., for high liquid flow rates B may deplete despite a low A solubility and a plentiful B supply in the bulk liquid because the inside tube wall is exposed directly to the bulk gas where the A supply rate is large. For the pellets high flows lead to complete wetting and no B depletion for a feed liquid with the same composition. Below we show that this is compensated for by the much more rapid A supply rate at high liquid flows obtained with the tube design.

The high flowrate behavior evident in Figure 6 for the $C_A{}^o = 0.01$ case is dictated by the virtual absence of A in the entering liquid. As the flowrate of the unsaturated liquid increases the conentration increase of A in the liquid along the tube length decreases. For very large flowrates the tube shell side is exposed to a liquid with a composition equal to its inlet composition over the entire tube length. Thus, with an increasing flowrate the efficiency decreases simply because the outer tube surface is exposed to a vitually unsaturated liquid over an increasing fraction of the tube length. Since the concentration of B in the flowing liquid also is close to its inlet value under these conditions, the pseudo first-order rate constant is essentially constant over the entire tube length. A point to note is that the efficiency approaches an asymptotic value that is much larger than the inlet liquid degree of saturation. On the other hand, for the pellet string the lower bound on the efficiency at high flowrates is necessarily the inlet liquid degree of saturation since the pellets are fully wetted (Figure 2). As we point out later, this is a major advantage of the tube design.

At low liquid flowrates the behavior is dictated by a different effect. For the inactive catalyst the efficiency approaches unity since the low flowrate coupled with a negligible reaction rate enables the liquid to become fully saturated with A. It should be noted that as the flowrate closely approaches zero the efficiency eventually decreases to zero. The reason for this is more apparent for the more active tubes. For the sufficiently active tube the concentration of B decreases considerably along the tube length. This decreases the pseudo first-order rate constant $k(z) = k' c_{Bf}(r = R_2, z)$. As a result, E decreases as the flowrate decreases. The trends at high and low flowrates are evident for the ($\phi_o = 1$, $C_A{}^o = 0.01$) case in which the efficiency exhibits a maximum at an intermediate flowrate.

WELL-STIRRED LIQUID CATALYTIC TUBE REACTOR

Another possible configuration for the catalytic tube reactor is gas flow through the tube core as before, but agitated liquid flow in the annular gap. The liquid enters and leaves through two or more ports. With sufficient agitation the bulk liquid may be considered well mixed, and thus the external tube surface is exposed to a uniform environment.

The model of this reactor is identical to the cocurrent flow reactor with regards to the gas core and intratube balances, Eqns. (C.1) - (C.5). However, the mass balances in the bulk liquid as well as the boundary conditions at the bulk liquid - tube surface interface ($r = R_2$) are different. Eqns. (C.6) and (C.7) are bulk liquid balances for species A and B, respectively. Each equation balances the net rate of material loss (first term) with the flux intergrated over the external tube surface (second term). The new boundary conditions (C.10) and (C.11) balance the external transport flux with the intraparticle flux at the external tube surface, at each point along the tube length. Thus, a boundary layer transport resistance exists for finite values of the Sherwood numbers, Sh_A and Sh_B. Their values depend on the nature of the flow. An approximate lower bound for Sh_A and Sh_B can be obtained by assuming that the liquid flows normal to the tube without recirculation. This corresponds to multiple injection points along the tube length without agitation. Mass transfer correlations are available for this situation, Eqns. (C.12) - (C.13). However, in keeping with the well-stirred assumption (with agitation), it is expected that Sh_A and Sh_B are higher than the values predicted by these correlations.

Solution of this model requires an iterative semi-implicit forward integration algorithm (described in more detail elsewhere; Cini and Harold, 1989). Briefly, integration of Eqns.(C.1) - (C.3) requires an initial estimate for the bulk liquid concentrations, C_{Ab} and C_{Bb}. A solution is obtained when the integration along the tube length satisfies the integral balances (C.6) and (C.7). Expressions for the species B fractional conversion (X_B) and the efficiency (E) are given by Equations (C.14) and (C.15), respectively. Efficiency calculations with this model will be presented elsewhere (Cini and Harold, 1989).

PERFORMANCE COMPARISON

In this section the performance of the string of pellets (SP), cocurrent flow catalytic tube (CF), and the well-stirred liquid catalytic tube (WS) reactors are compared. The question to be addressed is the following: Given a fixed catalyst volume and reactor length, which design performs best and under what conditions? Considering the large number of parameters it is impractical to probe all the possible conditions. Instead, a set of typical values of physical properties, transport, and kinetic parameters are assigned, as provided in Table 4.

The approach is to insert the string of spherical pellets of radius R_p or catalytic tube of inner and outer radius R_1, and R_2 into a shell of radius R_s and length L (see Figures 1.a and 5). The fixed catalyst volume is $(4 \pi R_p{}^3/3)$ $(L/2 R_p)$ for the pellet string and $\pi(R_2{}^2 - R_1{}^2)L$ for the tube. A string of 50 pellets with radius $R_p = 0.25$ cm is selected as the prototype trickle-bed (L=25 cm, V_c= 3.272 cm^3). Two tubes (CF1 and CF2) of length 25 cm are used with wall thicknesses of 0.10 cm and 0.05 cm, respectively. We should point out that these wall thicknesses give a reference diffusion length, $(R_2 - R_1)/2$, which is less than the diffusion length for the spheres, $R_p/3$. This point is considered below. The main diagnostic used in the comparison is the overall rate per unit volume, $E\,(k' c_{Bf}\, c_{AG}/H)_{max}$, where the term in brackets is the maximum rate attainable; i.e., the rate obtained with no intraparticle or external transport limitations, or axial concentration gradients. The rate is then calculated as a function of the volumetric liquid flow rate. The catalytic activity, degree of saturation of the liquid by the volatile reactant A, and the feed concentration of nonvolatile reactant B are the operating parameters whose impact is to be checked.

The string of pellets is first compared with the cocurrent flow catalytic tubes of wall thickness 0.1 and 0.05 cm. The impact of the catalytic activity (k') on the overall rate dependence on volumetric liquid flow rate (q_f) is shown in Figure 7. The string of pellets curves are solid and the curves for the catalytic tube of thickness 0.1 (0.05) cm are dashed (dotted). The feed liquid is completely saturated with volatile A and the feed concentration of nonvolatile B is 10^{-3} gmole/cm^3. The flowrate is varied between 0 and up to 20 cc/sec. For the liquid properties and sphere size selected, complete wetting of the pellets as predicted by Eqn. (2) occurs for $q_f^o = 0.18$ cm /sec. The maximum attainable rate is 1.89×10^{-8} k' gmole A/cm^3 sec.

Each reactor individually exhibits the expected trends. The string of pellet rate has a very slight local rate maximum in the partial wetting regime. Otherwise, the rate is an increasing function of flowrate. In the fully wetted regime, the rate is less sensitive to q_f for the catalysts of low activity. This is expected since external transport is not the limiting process. For a fixed q_f the rate increases with k'. The rate is proportional to $(k')^a$, where $a < 0.5$, at high flowrates and for the more active catalyst. This indicates that external transport is not insignificant for such conditions. The tube rate is a monotonic increasing function of q_f and, as in the case of the pellet string, is more sensitive to q_f for the more active catalyst.

Figure 7 demonstrates that the catalytic tube of wall thickness 0.05 cm (CF2) is the most efficient over the entire flow rate range and for each activity tested. The tube of thickness 0.1 cm (CF1) is the second best and the pellet string (PS) is the poorest performer. For example, for the k' = 100 cc/(gmole)(sec) and $q_f = 5$ cm^3/sec, the rates scale approximately 3 : 1.5 : 1. The gap between CF1 and PS widens as k' increases.

The shorter intraparticle diffusion length is the reason for the improvements exhibited by the tubes. If we demand that the tube wall thickness satisfy $(R_2 - R_1)/2 = R_p/3$, the resulting calculations indicate that the string of pellets perform marginally better with all other conditions fixed (Table 4). This suggests that intraparticle diffusion is the controlling rate process for the catalytic activities considered in Figure 7. Below we examine the outcome for activities that are more representative (i.e., higher) of noble metal catalyzed hydrogenation reaction systems.

In Figure 8 the advantage of the catalytic tube design is especially apparent for the case of a liquid containing only 1% of its equilibrium allotment of volatile A. As noted earlier, the completely wetted pellets perform poorly since the feed liquid film, containing little volatile reactant, acts as a barrier to species A supply along the length of the pellet string. This effect dominates if the contact time of a liquid element on the pellet string is sufficiently short compared to the characteristic time for A to diffuse through the liquid film from the bulk gas. The catalytic tube does not suffer from this limiation because species A can enter the pores directly from the gas filled core. Figure 8 reveals at least an order of magnitude rate improvement with the tube design.

Although the previous two figures demonstrate the advantage of the cocurrent flow catalytic tube, Figure 9 demonstrates a limitation. This occurs if the nonvolatile reactant B is in shorter supply. If the B feed concentration is decreased by an order of magnitude to 10^{-4} mole/cm^3 the inlet apparent activity, $k(z = 0) = k' c_{Bf}(z = 0)$, is as well, for a fixed intrinsic activity k'. For the completely wetted pellets $c_{Bf}(z = 0)$ is still sufficiently larger than $c_{AG}(z=0)/H$ to insure no depletion of B within the pellets. A comparison of Figures 7 and 9 reveals that for the same apparent inlet activity the string of pellets gives essentially the same overall rate. However, this is not the case for the catalytic tube. The facts that the tubes are wetted only on the outside surface, and that the rate of volatile A supply is very high from the gas-filled core results in a depletion of B within the tube wall. For example, depletion does not occur in the thinnest tube (CF2) for $k(z = 0) = 0.1$ sec^{-1} and $c_{Bf}(z = 0) = 10^{-3}$ gmole/cm^3. The overall rate for $q_f = 5$ cm^3/sec is about 5.7×10^{-7} gmole/(cm^3)(sec) (Figure 7). However, for $k(z = 0) = 0.1$ sec^{-1} and $c_{Bf}(z=0) = 10^{-4}$ gmole/cm^3 depletion occurs in an inner shell along the tube length. The overall rate for $q_f = 5$ cm^3/sec is about 4.8×10^{-7} gmole/(cm^3)(sec) (Fig. 9). This depletion effect is more severe as the activity increases.

Except for the case of a short supply of nonvolatile reactant the pellet string is out-performed by the cocurrent flow catalytic tube. The next question to address is whether the well stirred or cocurrent flow is the better design. Figure 10 shows a plot of the overall rate as a function of the liquid flowrate for three activities. The solid curves are the pellet string results and the dashed ones denote the well stirred liquid catalytic tube. Values for Sh_A and Sh_B are calculated according to $Sh_A = 100\, q_f^{1/3}$, $Sh_B = 2^{0.3}\, Sh_A$. For $q_f > 1$ these values are sufficiently large that the liquid-to-solid external mass transport is effectively eliminated. The well stirred liquid catalytic tube is clearly superior to the pellet string, as indicated by the rather large increase in overall rate for a given flowrate and activity. In fact, the improvement with the well stirred design exceeds that of the cocurrent flow design with the same catalytic tube (compare Figures 10 and 7).

The improvement exhibited by the catalytic tube is especially apparent for very active catalysts under conditions in which external mass transport is the primary controlling process. Suppose one has the choice between impregnating a single spherical pellet with a thin surface shell of a catalytic material, or a tubular support with a thin surface shell on its inner wall. If one demands that the total active catalyst volume is equal for both catalysts, and that the active zone and overall diffusion lengths are equal for both, the following expressions are derived which relate the dimensions of the sphere of radius R_p and active zone thickness δ_c to the tube of length $2R_p$, inner radius R_1, and outer radius R_2 (see Figure 11):

$$R_1 = R_p/2 - \delta_c \quad (4.a)$$

$$R_c = R_p/2 + \delta_c \quad (4.b)$$

$$R_2 = 7R_p/6 - \delta_c \quad (4.c)$$

$R_c - R_1$ is the thickness of the active zone in the tube.

It is now instructive to compare the active zone effectiveness obtained for the fully wetted sphere to that for the tube exposed to liquid flow on the shell side and gas flow in the core. It is assumed that the nonvolatile reactant is in large excess and that gradients in the direction of flow are negligible. The models and corresponding solutions are easily derived and will be presented elsewhere (Cini and Harold, 1989).

Figure 12 shows a plot of the active zone effectiveness as a function of the Thiele modulus based on the overall diffusion length for both the sphere and the tube. An active layer of thickness 2% of the pellet radius is considered (i.e., $s_c = 1 - \delta_c/R_p = 0.98$). The tube dimensions are given by Eqns. (4.a - c). For each catalyst the dimensionless overall external mass transport coefficient, $Sh = (V_p/S_x)\, k_c/D_{Ae}$, is varied between unity and 1000. For the tube, k_c represents the transport of the volatile reactant from the bulk liquid; it is assumed that the gas side transport resistance is negligible. The effectiveness is essentially independent of the liquid side external transport for this reason. On the other hand, the fully wetted pellet effectiveness is a strong decreasing function of the overall Sherwood number since the volatile reactant is only supplied via the liquid. The calculations clearly show the enhanced catalyst utilization obtained with the tube. We should note that typical Sherwood number values for a fully wetted pellet are in the range 1 - 100. The Sh = 100 calculations reveal that the tube performs better than the sphere for all catalytic activities (i.e., Thiele moduli) exceeding a critical value. For $Sh \leq 10$ the tube performs better for virtually the entire Thiele modulus range considered. These calculations definitely demonstrate the true superiority of the tube for very active catalysts.

CONCLUDING REMARKS

In this study we have confronted the question of whether or not the catalytically impregnated ceramic tube is an attractive alternative to the cocurrent downflow trickle-bed reactor for carrying out a catalytic reaction between volatile and nonvolatile compounds. We have developed sufficiently detailed mathematical models of (i) a string of suspended spherical pellets contacted by cocurrent gas-liquid flow, which we consider to be the prototype trickle-bed reactor, of (ii) the catalytic tube with cocurrent flow of gas in the tube core and liquid in the shell side, and of (iii) the same catalytic tube immersed in an agitated liquid.

The simulations reveal the benefits and limitations of contacting the gas and liquid phases on opposite sides of a hollow tubular catalytic wall. The main advantage is a result of the direct supply of the volatile reactant into the catalyst pores from the gas, thereby avoiding the liquid film mass transfer resistance. This design feature improves the overall catalyst performance for very active catalysts. For moderately active catalysts the simulations reveal the need to make the tube wall as thin as possible, thereby reducing the intraparticle transport resistance. One does not have as much flexibility in reducing the pellet size in the trickle-bed because of pressure drop considerations. Finally, the segregated flow design advantage is readily apparent if the feed liquid is not saturated with the volatile reactant.

Now that the benefits of the tubular catalyst design have been exhibited, the next more challenging step is to make the tubular multiphase catalysts and to demonstate the performance improvements in the laboratory. We have proceeded in this direction, and will present our progress in future papers.

NOTATION

a	volume fraction of active shell of spherical pellet
a_A	dimensionless parameter defined in Appendices A, B, C
a_I	dimensionless parameter defined in Appendices A, B, C
A	volatile reactant
b_A	dimensionless parameter defined in Appendices A, B, C
b_I	dimensionless parameter defined in Appendices A, B, C
B	nonvolatile reactant
Bi_i	overall Biot nember for species i
Bi_i^o	external part of Biot number for species i
Bi_{gs}^*	defined by Eqn. (A.12)
c_i'	concentration of species i within tube

c_{if}	concentration of species i in liquid film over pellet string reactor or liquid film over pellet string reactor liquid in cocurrent flow reactor
c_{ib}	concentration of species i in bulk liquid within well stirred reactor
c_{ib}^o	concentration of species i in inlet liquid to well stirred reactor
c_{iG}	concentration of species i in gas
C_i'	dimensionless intratube concentrations of species i (Appendices B and C)
C_{if}	$c_{if}/c_{if}(z=0)$
C_{iG}	$c_{iG}/c_{iG}(z=0)$
C_{ib}	c_{ib}/c_{ib}^o
C_A^o	degree of feed liquid saturation, $c_{Af}(z0)/(c_{AG}(z=0)/H$
C_f^o	ratio of inlet liquid B concentration to equilibrium solubility of A at inlet
D_i	bulk liquid phase diffusivity of species i
D_{ie}	intraparticle effective diffusivity of i
E	efficiency of pellet string or catalytic tube
E_w	degree of wetting (wetting efficiency)
g	acceleration due to gravity
H	Henry's law constant
H'	Heaviside function
I	inert species in gas
k	pseudo first-order rate constant
k'	bimolecular rate constant
k_{ci}	external mass transport coefficient for species i for well-stirred reactor
k_{gl}	gas to liquid mass transport coefficient for species A
k_{lsi}^o	external liquid-to-solid mass transport coefficient for species i
k_{lsi}	overall liquid-to-solid mass transport coefficient for species i
k_{gs}^o	external gas-to-solid mass transport coefficient of A
k_{gs}	overall gas-to-solid mass transport coefficient of A
L	length of pellet string or catalytic tube
m	parameter defined in Appendix A
m_o	parameter defined in Appendices B and C
M_i	molecular weight of species i
n	number of collocation points
N	number of pellets
P_f, P_G	parameters defined in Appendix A
P_t	parameters defined by Eqn. (B.15)
Pe_G	parameter defined in Appendices B and C
Pe_{fi}	Peclet number of species i (Appendix B)
q_f	volumetric liquid flowrate
q_f^o	critical flowrate below which partial setting occurs
Q	q_f/q_f^o
Q_A,Q_B	parameters defined in Appendix C
r	radial coordinate
r*	radial location of B depletion
R	gas constant
R_1, R_2	inner and outer radius of tube
R_c	distance from tube center to edge of active layer
R_p	radius of pellet
R_s	radius of reactor shell
Re_f	Reynolds number for normal flow over tube (Appendix C)
s	dimensionless lateral (x) distance in slab-type pellet
s_c	fraction of radius occupied by inert core in surface shell impregnated sphere
s*	lateral location of B depletion in pellet
S	perimeter of pellet (= $2\pi R_p$)
Sc_i	Schmidt number for i
S_x	external pellet area
Sh_i	Sherwood number of species i (Appendix C)
u_{if}	dimensionless liquid film concentration of species i (Appendix A)
u_{iw}	dimnsionless intraparticle concentration of i in wetted part (Appendix A)
u_{in}	dimensionless intraparticle concentration of i in nonwetted part with B present (Appendix A)
U	dimensionless velocity
v_f	velocity of liquid in cocurrent flow reactor (Appendix B)
v_G	velocity of gas
V_p	pellet volume
w	proportionality constant in Hartley-Murgatroyd criterion (Eqn.(2))
x	lateral distance coordinate in pellet
x^*	lateral location of B depletion
x_i	mole fraction of species i in gas phase
X_i	fractional conversion or species i
y	coordinate normal to wetted pellet surface
z	coordinate in flow direction
Z	z/L

Greek Letters

α	aspect ratio of slab catalyst (Appendix A)
β_i	parameter defined in Eqns. (A.20-A.21)
β_{oi}	parameter defined by Eqns. (A.22-A.23)
γ	parameter defined in appendices
Γ	mass flow rate per unit pellet width for single pellet model; parameter in cocurrent flow model (Appendix B)
Γ_o	minimal Γ below which partial wetting occurs
δ	V_p/S_x
δ_c	thickness of active layer on surface shell impregnated sphere
Δ	D_{Be}/D_{Ae}
Δ	D_{ie}/D_i
ε	parameter defined in Appendix A
ε_p	particle porosity
κ	k^o_{gs}/k_{gl}
λ	R_2/R_s
λ_p/λ_s	parameters defined in Appendix A
Λ	parameter defined in Appendix A

μ_L, μ_G	liquid and gas viscosity
ν	stoichiometric coefficient
ξ	r/R_s
ξ	$2r/(R_2-R_1)$
ξ^*	dimensionless location of B depletion
ξ_1, ξ_2	R_1/R_s, R_2/R_s
ξ_p	R_p/R_s
ρ_f	liquid density
σ	interfacial tension
ϕ	Thiele modulus based on concentration of B
ϕ_o	Thiele modulus based on inlet concentration of B
θ	angle from vertical

Subscripts

act	active
A	species A
b	bulk
B	species B
c	inactive core, or pertaining to k_{ci}
d	species B depleted zone in pellet
e	effective or equilibrium
f	film or liquid
g,G	gas
I	species i
l	liquid
o	reference
p	pellet
s	shell
1	radial position of inner tube wall
2	radial position of outer tube wall

Superscripts

o	reference
^	average
*	depletion point

LITERATURE CITED

Akyurtlu, J.F., A.A. Akyurtlu, and C.E. Hamrin, Chem. Eng. Sci., 40, 1785 (1985).

Akyurtlu, J.F., A.A. Akyurtlu, and C.E. Hamrin, Chem. Eng. Comm., 66, 169 (1988).

Cini, P. and M.P. Harold, to be submitted (1989).

de Vos, R., and C.E. Hamrin, Jr., Chem. Eng. Sci., 37, 1711 (1982).

de Vos, R., V. Hatziantoniou, and N.-H. Schoon, Chem. Eng. Sci., 37, 1719 (1982).

Germain, A., Crine, P. Marchot, and G.A. L'Homme, Am. Chem. Socl Symp. Ser., 65, 411 (1978).

Harold, M.P., and K.M. Ng, AIChE J., 33, 9 (1987).

Hartley, D.E., and W. Murgatroyd, Int. J. Heat and Mass Transfer, 7, 1003 (1964).

Hatziantoniou, V., and B. Andersson, Ind. Eng. Chem. Fund., 21, 451 (1982).

Hatziantoniou, V., and B. Andersson, Ind. Eng. Chem. Fund., 23, 82 (1984).

Hatziantoniou, V., B. Andersson,and N.-H Schoon, Ind. Eng. Chem. Process Des. Dev., 25, 964 (1986).

Hatziantoniou, V., B. Andersson, T. Larsson, N.-H. School, L. Carlsson, S. Schwartz, and K.-B. Wideen, Ind. Eng. Chem. Process Des. Devel., 25, 143 (1986).

Herskowitz, M., R.G. Carbonell, and J.M. Smith, AIChE J., 25, 272 (1979).

Lynn, S., J.R. Straatemeier, and H. Kramers, Chem. Eng.,4, 63 (1952).

Morita, S. and J.M.Smith, Ind. Eng. Chem. Fund., 17, 113 (1978).

Norman, W.S., and V. McIntyre, Trans. Inst. Chem. Eng., 38, 301 (1960).

Norris, R.H. and D.D. Streid, Trans. A.S.M.E., 36, 525 (1940).

Ng, K.M. AIChE J., 32, 115 (1986).

Satterfield, C.N., A.A. Pelossof, and T.K. Sherwood, 15, 226 (1969).

Satterfield, C.N., and F. Ozel, Ind. Eng. Chem. Fundam., 21, 451 (1977).

Sato, Y.,T. Hirose, F. Takahashi, M. Toda, and Y. Hashiguchi, J. Chem. Eng. Japan, 6, 315 (1973).

Tarhan, M.O., Catalytic Reactor Design, McGraw-Hill Book Company, New York (1983).

Thomas, J. Chem. Soc., Part II, 573 (1946).

Villadsen, J. and M.L. Michelsen, Solution of Differential Equation Models by Polynomial Approximation, Princeton-Hall, Inc., Englewood Cliffs, New Jersey (1978).

Yentekakis, I.V., and C.G. Vayenas, Chem. Eng. Sci., 42, 1323 (1987).

Appendix A: String of pellets model for case of B depletion in nonwetted part of the pellet.

Governing equations:

Bulk gas ($\xi_p < \xi < 1$):

$$a_I\, C_{IG}\, U \frac{dU}{dZ} + a_A\, C_{AG}\, U \frac{dU}{dZ} + b_I \frac{dC_{IG}}{dZ} + b_A \frac{dC_{AG}}{dZ} + U = 0 \qquad \text{(A.1)}$$

$$\frac{d}{dZ}(UC_{IG}) = 0 \qquad \text{(A.2)}$$

$$\frac{d}{dZ}(UC_{AG}) + \frac{E_w\, Bi_{gl}\, \lambda_s}{P_G} C_{AG}(1 - u_{Af})$$

$$+ \frac{Bi_{gs}\, \lambda_s}{P_G} C_{AG} \int_{E_w}^{s^*} (1 - u_{An})\, ds$$

$$+ \frac{Bi^*_{gs}\, \lambda_s}{P_G} C_{AG} \int_{s^*}^{1} (1 - u_{An,d})\, ds = 0 \qquad \text{(A.3)}$$

Liquid film:

$$\frac{dC_{Af}}{dZ} + \frac{E_w\, Bi_{gl}\, \lambda_p}{P_f} \frac{C_{AG}}{C_A^o} (u_{Af} - 1)$$

$$+ \frac{Bi_{lsA}\, \lambda_p}{P_f} \frac{C_{AG}}{C_A^o} \int_o^{E_w} (u_{Af} - u_{Aw})\, ds = 0 \qquad \text{(A.4)}$$

$$\frac{d\,C_{Bf}}{dZ} + \frac{\Delta\, Bi_{lsB}\, \lambda_p}{P_f} \frac{C_{AG}}{C_A^o\, C_f^o} \int_0^{E_w} (u_{Bf} - u_{Bw})\, ds = 0 \qquad \text{(A.5)}$$

Pellet:

Wetted part ($0 < s < E_w$):

$$\frac{d^2u_{Aw}}{ds^2} + \alpha\, Bi_{lsA}\,(u_{Af} - u_{Aw}) - \phi^2\, u_{Aw} = 0 \tag{A.6}$$

$$\frac{d^2u_{Aw}}{ds^2} + \alpha\, Bi_{lsB}\,(u_{Bf} - u_{Bw}) - m\,\phi^2\, u_{Aw} = 0 \tag{A.7}$$

Nonwetted part with B present ($E_w < s < s^*$):

$$\frac{d^2u_{Aw}}{ds^2} + \alpha\, Bi_{gs}\,(1 - u_{An}) - \phi^2\, u_{An} = 0 \tag{A.8}$$

$$\frac{d^2u_{Bn}}{ds^2} - m\,\phi^2 u_{An} = 0 \tag{A.9}$$

Nonwetted part with B depleted ($s^* < s < 1$):

$$\frac{d^2u_{An,d}}{ds^2} + \alpha\, Bi^*_{gs}\,(1 - u_{An,d}) = 0 \tag{A.10}$$

$$u_{Bn,d} = 0 \tag{A.11}$$

$$Bi^*_{gs} = (Bi_{gs} - Bi^o_{gs})\, s^* + Bi^o_{gs} \tag{A.12}$$

$$\frac{1}{Bi_{gs}} = \frac{D_{Ae}}{\pi\, R_p\, k_{gs}} = \frac{1}{Bi^o_{gs}} + \frac{\varepsilon}{\phi} \tag{A.13}$$

$$\frac{1}{Bi_{lsA}} = \frac{D_{Ae}}{\pi\, R_p\, k_{lsA}} = \frac{1}{Bi^o_{lsA}} + \frac{\varepsilon}{\phi} \tag{A.14}$$

$$\frac{1}{Bi_{lsB}} = \frac{D_{Be}}{\pi\, R_p\, k_{lsB}} = \frac{1}{Bi_{lsB}} + \frac{\varepsilon}{\phi} \tag{A.15}$$

$$\varepsilon = \frac{1}{\tanh(\phi/\alpha)} - \frac{\alpha}{\phi} \tag{A.16}$$

$$Bi^{o}_{lsA} = 1.527\,\Lambda\,(\Gamma/\Gamma_o)\left(\frac{1}{\beta_A}\right)^{-2/3} \tag{A.17}$$

$$Bi^{o}_{lsB} = 1.527\,\Lambda\,\Delta^{-1}\,(\Gamma/\Gamma_o)\left(\frac{1}{\beta_B}\right)^{-2/3} \tag{A.18}$$

$$Bi_{gl} = 1.695\,\Lambda\,(\Gamma/\Gamma_o)\left(\frac{1}{\beta_A}\right)^{-1/2} \tag{A.19}$$

$$\frac{1}{\beta_A} = \frac{1}{\beta_{oA}}, \quad \frac{1}{\beta_B} = \frac{1}{\beta_{oB}} \quad \text{if } E_w < 1 \tag{A.20}$$

$$\frac{1}{\beta_A} = \frac{1}{\beta_{oA}}\left(\frac{\Gamma}{\Gamma_o}\right)^{4/3}, \quad \frac{1}{\beta_B} = \frac{1}{\beta_{oB}}\left(\frac{\Gamma}{\Gamma_o}\right)^{4/3} \quad \text{if } E_W = 1 \tag{A.21}$$

$$\frac{1}{\beta_{oA}} = \frac{1}{2.58 D_A}\left(\frac{81\,\nu\,q_o^4}{128\,\pi^4\,g\,R_p^7\,\sin^4\theta}\right)^{1/3} \tag{A.22}$$

$$\frac{1}{\beta_{oB}} = \Delta^{-1}\left(\frac{1}{\beta_{oB}}\right) \tag{A.23}$$

Boundary conditions

Inlet bulk gas ($\xi_p < \xi < 1$, $Z = 0$):

$$C_{IG}(\xi, Z=0) = 1 \qquad C_{AG}(\xi, Z=0) = 1 \qquad U(\xi, Z=0) = 1 \tag{A.24.a–c}$$

Inlet liquid film ($z = 0$):

$$C_{Af}(Z=0) = 1 \qquad C_{Bf}(Z=0) = 1 \tag{A.25.a–b}$$

Intrapellet:

Middle of wetted part ($s = 0, Z$):

$$\frac{du_{Aw}}{ds} = \frac{du_{Bw}}{ds} = 0 \quad \text{at} \quad s = 0 \tag{A.26. a–b}$$

Wetted-nonwetted boundary: $(s = E_w, Z)$:

$$u_{Aw} = u_{An} \qquad \frac{du_{Aw}}{ds} = \frac{du_{An}}{ds} \tag{A.27.a–b}$$

$$u_{Bw} = u_{Bn} \qquad \frac{du_{Bw}}{ds} = \frac{du_{Bn}}{ds} \tag{A.28.a–b}$$

Point of nonvolatile reactant B depletion $(s = s^*, Z)$:

$$u_{An} = u_{An,d} \qquad \frac{du_{An}}{ds} = \frac{du_{An,d}}{ds} \tag{A.29. a–b}$$

$$u_{Bn} = 0 \qquad \frac{du_{Bn}}{ds} = 0 \tag{A.30.a–b}$$

Middle of nonwetted part $(s = 1, Z)$:

$$\frac{du_{An,d}}{ds} = 0 \tag{A.31.a–b}$$

Efficiency and Conversion Expressions

Fractional conversion of A:

$$X_A = \frac{1}{1+P} [\, 1 - C_{AG}(Z=1)\, U(Z=1) \; + \; P\,(1 - C_{Af}(Z=1)]$$

$$\text{where} \quad P = \frac{P_f \lambda_s C_A^o}{P_G \lambda_p} \tag{A.32}$$

Fractional conversion of B:

$$X_B = 1 - C_{Bf}\,(Z=1) \tag{A.33}$$

$$E = \frac{C_f^o C_A^o P_f \alpha X_B}{\nu \lambda_p \phi_o^2} \tag{A.34}$$

Surface shell impregnated pellets:

$$E = a \frac{C_f^o C_A^o P_f \alpha X_B}{\nu \lambda_p \phi_o^2} \tag{A.35}$$

Independent variables

$$Z = z/L \qquad \xi = r/R_s \qquad S = x/(S/2)$$

State variables

$$C_{AG} = \frac{c_{AG}(z)}{c_{AG}(z=0)} \qquad C_{IG} = \frac{c_{IG}(z)}{c_{IG}(z=0)} \qquad C_{Af} = \frac{c_{Af}(z)}{c_{Af}(z=0)}$$

$$C_{Bf} = \frac{c_{Bf}(z)}{c_{Bf}(z=0)} \qquad U = \frac{\hat{v}_G(z)}{\hat{v}_G(z=0)} \qquad u_{Aw} = \frac{c'_{Aw}(x, z)}{c_{AG}(z)/H}$$

$$u_{An} = \frac{c'_{An}(x, z)}{c_{AG}(z)/H} \qquad u_{An,d} = \frac{c'_{An,d}(x, z)}{c_{AG}(z)/H} \qquad u_{Bw} = \frac{c'_{Bw}(x, z)}{c_{AG}(z)/H}$$

$$u_{Bn} = \frac{c'_{Bn}(x, z)}{c_{AG}(z)/H} \qquad u_{Af} = \frac{c_{Af}(z)}{c_{AG}(z)/H} \qquad u_{Bf} = \frac{c_{Bf}(z)}{c_{AG}(z)/H}$$

Parameters

$$a_I = \frac{c_{IG}(z{=}0)\, M_I\, \hat{v}_G(z{=}0)}{\gamma L} \qquad a_A = \frac{c_{AG}(z{=}0)\, M_A\, \hat{v}_G(z{=}0)}{\gamma L}$$

$$b_I = \frac{R\,T\,c_{IG}(z{=}0)}{\gamma L\, \hat{v}_G(z{=}0)} \qquad b_A = \frac{R\,T\,c_{AG}(z{=}0)}{\gamma L\, \hat{v}_G(z{=}0)}$$

$$\gamma = \frac{8\mu_G}{R_p^2}\left[\frac{1-\xi_p^4}{1-\xi_p^4} - \frac{1-\xi_p^2}{\ln(1/\xi_p)}\right]^{-1} \qquad \xi_p = \frac{R_p}{R_s}$$

$$\Delta = \frac{D_{Be}}{D_{Ae}} \qquad C_A^o = \frac{c_{Af}(z=0)}{c_{AG}(z=0)/H} \qquad C_f^o = \frac{c_{Bf}(z=0)}{c_{Af}(z=0)}$$

$$P_G = \frac{R_S\,(1-\xi_p^2)\,\hat{v}_G(z=0)}{D_{Ae}} \qquad P_f = \frac{q_f}{\pi\, R_p\, D_{Ae}} \qquad \lambda_s = \frac{2\,L\,H}{\pi\, R_s}$$

$$\lambda_p = \frac{2L}{\pi R_p} \qquad \phi_o^2 = \left(\frac{S}{2}\right)^2 \frac{k' c_{Bf}(z=0)}{D_{Ae}} \qquad \phi^2 = \phi_o^2 C_{Bf}(z=0)$$

$$\alpha = \frac{S/2}{\delta} = \frac{\pi R_p}{V_p/S_x} = 3\pi \qquad m = \frac{\nu D_{Ae}}{D_{Be}}$$

$$Bi_{gl} = \frac{\pi R_p k_{gl}}{D_{Ae}} \qquad Bi_{gs}^o = \frac{\pi R_p k_{gs}^o \varepsilon_p}{D_{Ae}}$$

$$Bi_{lsA}^o = \frac{\pi R_p k_{lsA}^o \varepsilon_p}{D_{Ae}} \qquad Bi_{lsB}^o = \frac{\pi R_p k_{lsB}^o \varepsilon_p}{D_{Be}}$$

$$\Lambda = \frac{\varepsilon_p \Gamma_o}{\rho_f D_{Ae}} \qquad a = 1 - (R_{p,act}/R_p)^3$$

Appendix B: The catalytic tube with cocurrent gas-liquid flow for the case of no B depletion.

Governing equations

Bulk gas $(0 < \xi < \xi_1$ or $0 < \hat{\xi} < \hat{\xi}_1)$:

$$a_I C_{IG} U \frac{dU}{dZ} + a_A C_{AG} U \frac{dU}{dZ} + b_I \frac{dC_{IG}}{dZ} + b_A \frac{dC_{AG}}{dZ} + U = 0 \tag{B.1}$$

$$\frac{d}{dZ}(UC_{IG}) = 0 \tag{B.2}$$

$$Pe_G \frac{d}{dZ}(UC_{AG}) - C_{AG} \left.\frac{dC_A'}{d\hat{\xi}}\right|_{\hat{\xi}=\hat{\xi}_1} = 0 \tag{B.3}$$

Intratube $(\xi_1 < \xi < \xi_2$ or $\hat{\xi}_1 < \hat{\xi} < \hat{\xi}_2)$:

$$\frac{d^2C_A'}{d\hat{\xi}^2} + \frac{1}{\hat{\xi}}\frac{dC_A'}{d\hat{\xi}} - \phi_o^2 C_{Bf}(\hat{\xi}_2, Z)\, C_A' = 0 \tag{B.4}$$

$$\frac{d^2 C_B'}{d\hat{\xi}^2} + \frac{1}{\hat{\xi}}\frac{d C_B'}{d\hat{\xi}} - m_o\, \phi_o^2\, C_{AG}(Z)\, C_A' = 0 \tag{B.5}$$

Bulk liquid ($\xi_2 < \xi < 1$ or $\hat{\xi}_2 < \hat{\xi} < \hat{\xi}_s$):

$$P_{e_{fA}}\,\Gamma\left[1 - (\hat{\xi}/\hat{\xi}_S)^2 + \frac{1-\lambda^2}{\ln(1/\Gamma)}\ln(\hat{\xi}/\hat{\xi}_S)\right]\frac{\partial C_{Af}}{\partial Z} =$$

$$\frac{\partial C_{Af}}{\partial \hat{\xi}^2} + \frac{1}{\hat{\xi}}\frac{\partial C_{Af}}{\partial \hat{\xi}} + \left(\frac{R_2 - R_1}{2L}\right)^2 \frac{\partial^2 C_{Af}}{\partial Z^2} \tag{B.6}$$

$$Pe_{fB}\,\Gamma\left[1 - (\hat{\xi}/\hat{\xi}_S)^2 + \frac{1-\lambda^2}{\ln(1/\Gamma)}\ln(\hat{\xi}/\hat{\xi}_S)\right]\frac{\partial C_{Bf}}{\partial Z} =$$

$$\frac{\partial^2 C_{Bf}}{\partial \hat{\xi}^2} + \frac{1}{\hat{\xi}}\frac{\partial C_{Bf}}{\partial \hat{\xi}} + \left(\frac{R_2 - R_1}{2L}\right)^2 \frac{\partial^2 C_{Bf}}{\partial Z^2} \tag{B.7}$$

<u>Boundary conditions</u>

Inlet Bulk gas ($0 \le \hat{\xi} < \hat{\xi}_1$, $Z=0$):

$$C_{IG}(\hat{\xi}, Z=0) = 1 \qquad C_{AG}(\hat{\xi}, Z=0) = 1 \qquad U(\hat{\xi}, Z=0) = 1 \tag{B.8.a–c}$$

Inlet Bulk liquid ($\hat{\xi}_2 < \hat{\xi} < \hat{\xi}_S$, $Z=0$):

$$C_{Af}(\hat{\xi}, Z=0) = 1 \qquad\qquad C_{Bf}(\hat{\xi}, Z=0) = 1 \tag{B.9.a–b}$$

Boundary between bulk liquid and catalytic tube ($\hat{\xi} = \hat{\xi}_2$, $Z \ge 0$):

$$C_{AG}(Z)\,\Delta_A \frac{d C_A'}{d\hat{\xi}}\Big|_{\hat{\xi}=\hat{\xi}_2} = C_A^o \frac{d C_{Af}}{d\hat{\xi}}\Big|_{\hat{\xi}=\hat{\xi}_2} \tag{B.10}$$

$$C_{Bf}(\hat{\xi}_2, Z)\, \Delta_B \frac{d C_B'}{d\hat{\xi}}\Big|_{\hat{\xi}=\hat{\xi}_2} = \frac{dC_{Bf}}{d\hat{\xi}}\Big|_{\hat{\xi}=\hat{\xi}_2} \tag{B.11}$$

Boundary between bulk gas and catalytic tube $(\hat{\xi} = \hat{\xi}_f,\ Z \geq 0)$:

$$C_A'(\hat{\xi} = \hat{\xi}_1, Z) = 1 \qquad \frac{dC_B'}{d\hat{\xi}}\Big|_{\hat{\xi}=\hat{\xi}_1} = 0 \tag{B.12.a–b}$$

Boundary between bulk liquid and shell wall $(\hat{\xi} = \hat{\xi}_S,\ Z \geq 0)$:

$$\frac{dC_{Af}}{d\hat{\xi}}\Big|_{\hat{\xi}=\hat{\xi}_s} = 0 \qquad \frac{dC_{Bf}}{d\hat{\xi}}\Big|_{\hat{\xi}=\hat{\xi}_s} = 0 \tag{B.13.a–b}$$

Efficiency and conversion expressions

Fractional conversion of B:

$$X_B = 1 - \frac{2}{1+\lambda} \int_{\xi=\lambda}^{1} \frac{v_f(\xi)}{\hat{v}_f}\, C_{Bf}(Z=1, \xi)\, \xi\, d\xi$$

$$\frac{v_f(\xi)}{\hat{v}_f} = 2\Gamma\left[1 - \xi^2 + \frac{1-\lambda^2}{\ln(1/\lambda)} \ln(\xi)\right] \tag{B.14}$$

Fractional conversion of A:

$$X_A = \frac{1}{1+P_t}\Bigg[1 - C_{AG}(Z=1)\, U(Z=1) + P_t\left(1 - \frac{2}{1+\lambda}\int_{\xi=\lambda}^{1} \frac{v_f(\xi)}{\hat{v}_f}\, C_{Af}(Z=1,\xi)\, \xi\, d\xi\right)\Bigg]$$

$$P_t = \frac{Pe_{fA}'\, C_A^o\, (1+\lambda)}{4\, Pe_G\, \Delta_A\, \hat{\xi}_1\, (1-\lambda)} \tag{B.15}$$

Cocurrent flow catalytic tube efficiency:

$$E = \frac{Pe'_{fB}\,(1+\lambda)}{4\,(\hat{\xi}_1 + \hat{\xi}_2)\; m_o\, \phi_o^2\; \Delta_B\,(1-\lambda)}\, X_B \tag{B.16}$$

Independent variables

$$Z = z/L \qquad \xi = r/R_s \qquad \hat{\xi} = \frac{2r}{(R_2 - R_1)}$$

State variables

$$C_{AG} = \frac{c_{AG}(z)}{c_{AG}(z=0)} \qquad U = \frac{\hat{v}_G(z)}{\hat{v}_G(z=0)} \qquad C'_A = \frac{c'_A(r,z)}{c_{AG}(z)/H} \qquad C'_B = \frac{C'_B(r,z)}{C_{Bf}(r=R_2, z)}$$

$$C_{Af} = \frac{c_{Af}(r,z)}{c_{Af}(r, z=0)} \qquad C_{Bf} = \frac{c_{Bf}(r,z)}{c_{Bf}(r, z=0)} \qquad C_{IG} = \frac{c_{IG}(z)}{c_{IG}(z=0)}$$

Parameters

$$a_I = \frac{c_{IG}(z=0)\, M_I\, \hat{v}_G(z=0)}{\gamma\, L} \qquad a_A = \frac{c_{AG}(z=0)\, M_A\, \hat{v}_G(z=0)}{\gamma\, L}$$

$$b_I = \frac{RT\, c_{IG}(z=0)}{\gamma\, L\, \hat{v}_G(z=0)} \qquad b_A = \frac{RT\, c_{AG}(z=0)}{\gamma\, L} \qquad \gamma = \frac{8\mu_G}{R_1^2} \qquad \Delta_A = \frac{D_{Ae}}{D_A} \qquad \Delta_B = \frac{D_{Be}}{D_B}$$

$$Pe_G = \frac{R_1\,(R_2 - R_1)\,\hat{v}_G(z=0)H}{4\, D_{Ae}\, L} \qquad Pe_{fA} = \frac{(R_2 - R_1)^2\,\hat{v}_f}{2\, D_A\, L} \qquad Pe_{fB} = \frac{(R_2 - R_1)^2\,\hat{v}_f}{2\, D_B\, L}$$

$$Pe'_{fA} = (\hat{\xi}_s - \hat{\xi}_2)^2\, Pe_{fA} \qquad Pe'_{fB} = (\hat{\xi}_s - \hat{\xi}_2)^2\, Pe_{fB}$$

$$m_o = \frac{\nu\, D_{Ae}\, c_{AG}(z=0)/H}{D_{Be}\, c_{Bf}(z=0)} \qquad \phi_o^2 = \left(\frac{R_2 - R_1}{2}\right)^2 \frac{k'\, c_{Bf}(z=0)}{D_{Ae}}$$

$$\hat{\xi}_s = \frac{2\,R_s}{R_2 - R_1} \qquad \hat{\xi}_1 = \frac{2\,R_1}{R_2 - R_1} \qquad \hat{\xi}_2 = \frac{2\,R_2}{R_2 - R_1} \qquad \lambda = \frac{R_2}{R_s}$$

$$C_A^o = \frac{c_{Af}(z=0, r)}{c_{AG}(z=0)/H} \qquad \Gamma = \left[1 + \lambda^2 - \frac{1-\lambda^2}{\ln(1/\lambda)}\right]^{-1}$$

Appendix C: The well stirred liquid catalytic tube reactor.

Governing equations

Bulk gas ($0 < \xi < \xi_1$ or $0 < \hat{\xi} < \hat{\xi}_1$):

$$a_I\, C_{IG}\, U \frac{dU}{dZ} + a_A\, C_{AG}\, U \frac{dU}{dZ} + b_I \frac{dC_{IG}}{dZ} + b_A \frac{dC_{AG}}{dZ} + U = 0 \qquad \text{(C.1)}$$

$$\frac{d}{dZ}(U\, C_{IG}) = 0 \qquad \text{(C.2)}$$

$$Pe_G \frac{d}{dZ}(U\, C_{AG}) - C_{AG} \frac{d\,C_A'}{d\hat{\xi}}\Big|_{\hat{\xi} = \hat{\xi}_1} = 0 \qquad \text{(C.3)}$$

Intratube ($\xi_1 < \xi < \xi_2$ or $\hat{\xi}_1 < \hat{\xi} < \hat{\xi}_2$):

$$\frac{d^2 C_A'}{d\hat{\xi}^2} + \frac{1}{\hat{\xi}} \frac{dC_A'}{d\hat{\xi}} - \phi_o^2\, C_{Bb}\, C_A' = 0 \qquad \text{(C.4)}$$

$$\frac{d^2 C_B'}{d\hat{\xi}^2} + \frac{1}{\hat{\xi}} \frac{dC_B'}{d\hat{\xi}} - m_o\, \phi_o^2\, C_{AG}(Z)\, C_A' = 0 \qquad \text{(C.5)}$$

Bulk liquid ($\xi_2 < \xi < 1$ or $\hat{\xi}_2 < \hat{\xi} < \hat{\xi}_s$):

$$Q_A\,(C_A^o - C_{Ab}) - Sh_A \int_{Z=0}^{1} [C_{Ab} - C_{AG}(Z)\, C_A'(\hat{\xi}_2, Z)]\, dZ = 0 \qquad \text{(C.6)}$$

$$Q_B\,(1 - C_{Bb}) - Sh_B\, C_{Bb} \int_{Z=0}^{1} [1 - C_B'(\hat{\xi}_2, Z)]\, dZ = 0 \qquad \text{(C.7)}$$

Boundary conditions

Inlet bulk gas $(0 \le \hat{\xi} < \hat{\xi}_1, Z = 0)$:

$$C_{IG}(\hat{\xi}, Z=0) = 1 \qquad C_{AG}(\hat{\xi}, Z=0) = 1 \qquad U(\hat{\xi}, Z=0) = 1 \qquad \text{(C.8.a–c)}$$

Boundary between bulk gas and catalytic tube $(\hat{\xi} = \hat{\xi}_1, Z \ge 0)$:

$$C_A'(\hat{\xi} = \hat{\xi}_1, Z) = 1 \qquad \frac{dC_B'}{d\hat{\xi}}\Big|_{\hat{\xi} = \hat{\xi}_1} = 0 \qquad \text{(C.9.a–b)}$$

Boundary between bulk liquid and catalytic tube $(\hat{\xi} = \hat{\xi}_2, Z \ge 0)$:

$$Sh_A[C_{Ab}/C_{AG}(Z) - C_A'(\hat{\xi}_2, Z)] - \frac{dC_A'}{d\hat{\xi}}\Big|_{\hat{\xi} = \hat{\xi}_2} = 0 \qquad \text{(C.10)}$$

$$Sh_B[1 - C_B'(\hat{\xi}_2, Z)] - \frac{dC_B'}{d\hat{\xi}}\Big|_{\hat{\xi} = \hat{\xi}_2} = 0 \qquad \text{(C.11)}$$

$$Sh_A = Sc_A^{0.3}\,(0.35 + 0.34\,Re_f^{0.5} + 0.15\,Re_f^{0.58})\,\xi_2^{-1} \qquad \text{(C.12)}$$

$$Sh_B = Sh_A\left(\frac{Sc_B}{Sc_A}\right)^{0.3} \qquad \text{(C.13)}$$

Efficiency and conversion expressions

Fractional conversion of B:

$$X_B = 1 - C_{Bb} \qquad \text{(C.14)}$$

Well stirred liquid catalytic tube efficiency:

$$E = \frac{2}{\phi_o^2\,(\hat{\xi}_2^2 - \hat{\xi}_1^2)}\int_{Z=0}^{1}\left[\left(\hat{\xi}\,\frac{dC_A'}{d\hat{\xi}}\right)_{\hat{\xi} = \hat{\xi}_2} - \left(\hat{\xi}\,\frac{dC_A'}{d\hat{\xi}}\right)_{\hat{\xi} = \hat{\xi}_1}\right] dZ \qquad \text{(C.15)}$$

Independent Variables

$$Z = z / L \qquad \xi = r / R_s \qquad \hat{\xi} = \frac{2r}{(R_2 - R_1)}$$

State Variables

$$C_{AG} = \frac{c_{AG}(z)}{c_{AG}(z=0)} \qquad C_{IG} = \frac{c_{IG}(z)}{c_{IG}\ (z=0)} \qquad U = \frac{\hat{v}_G\ (z)}{\hat{v}_G(z=0)}$$

$$C'_A = \frac{c'_A(\, r, z\,)}{c_{AG}(z=0)/H} \qquad C'_B = \frac{c'_B(r, z\,)}{c_{Bb}} \qquad C_{Ab} = \frac{c_{Ab}}{c_{AG}(z=0)/H} \qquad C_{Bb} = \frac{c_{Bb}}{c^o_{Bb}}$$

Parameters

$$a_I = \frac{c_{IG}(z=0)\, M_I\, \hat{v}_G\ (z=0)}{\gamma\ L} \qquad a_A = \frac{c_{AG}(z=0)\, M_A\, \hat{v}_G\ (z=0)}{\gamma\ L}$$

$$b_I = \frac{RT\ c_{IG}(z=0)}{\gamma\ L\, \hat{v}_G\ (\, z=0)} \qquad b_A = \frac{RT\ c_{AG}(z=0)}{\gamma\ L} \qquad \gamma = \frac{8\mu_G}{R_1^2} \qquad Re_f = \frac{\rho_f\, q_f}{\pi L \mu_f}$$

$$Sh_A = \frac{(R_2 - R_1) k_{cA} \varepsilon_p}{2D_{Ae}} \qquad Sh_B = \frac{(R_2 - R_1) k_{cB} \varepsilon_p}{2D_{Be}} \qquad Sc_A = \frac{\mu_f}{\rho_f\, D_A} \qquad Sc_B = \frac{\mu_f}{\rho_f\, D_B}$$

$$Pe_G = \frac{R_1\ (R_2 - R_1\ )\ \hat{v}_G\ (z=0)H}{4\, D_{Ae}\ L} \qquad Q_A = \frac{q_f\ (R_2 - R_1)}{4\pi\, D_{Ae}\, R_2\, L} \qquad Q_B = \frac{q_f\ (R_2 - R_1)}{4\pi\, D_{Be}\, R_2\, L}$$

$$\phi_o^2 = \left(\frac{R_2 - R_1}{2}\right)^2 \frac{k'\, c^o_{bB}}{D_{Ae}} \qquad m_o = \frac{\nu\, D_{Ae}\, C_{AG}(z=0)/H}{D_{Be}\, c^o_{Bb}}$$

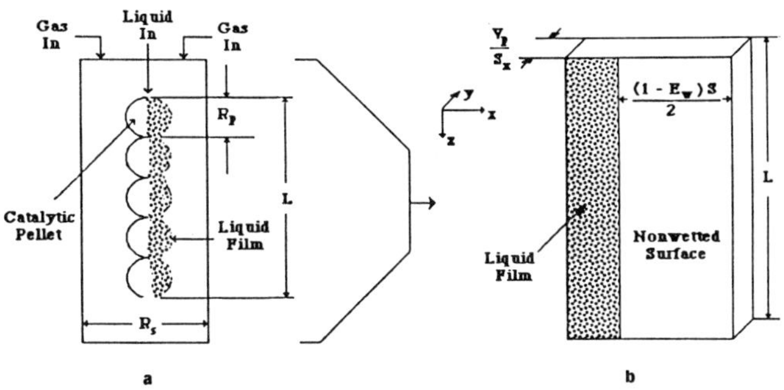

Figure 1. (a) is a schematic of the string of suspended spherical catalytic pellets contacted cocurrently by the gas and a liquid film. (b) shows the catalytic slab model.

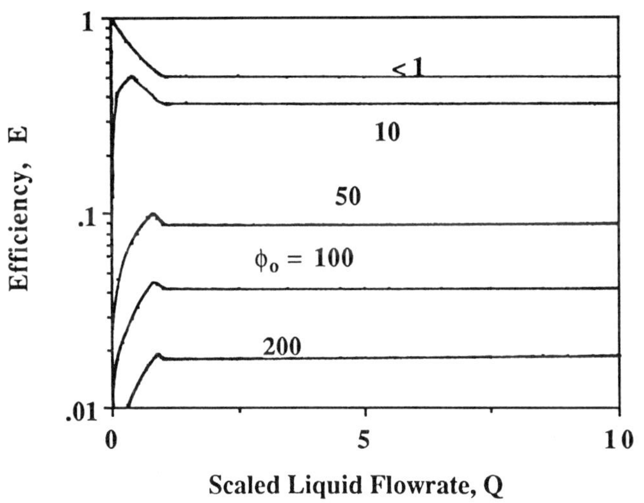

Figure 2. Efficiency of the pellet string as a function of the scaled liquid flowrate, $Q = q_f/q_f^0$, for several values of the catalytic activity (ϕ_0). The feed liquid is half-saturated with the volatile reactant A. Other parameter values appear in Table 1.

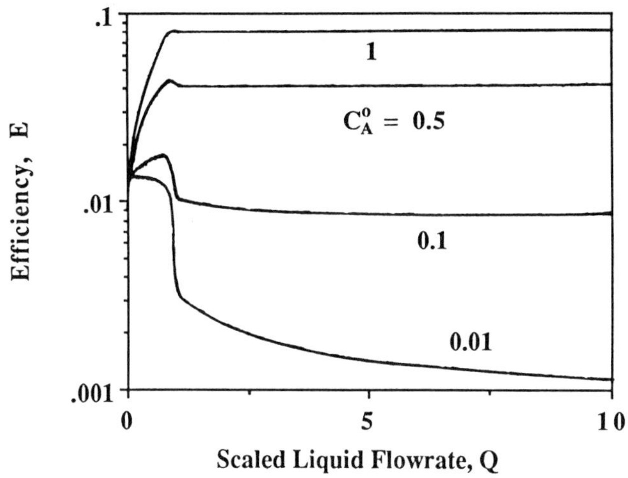

Figure 3. Efficiency of the pellet string as a function of the scaled liquid flowrate, $Q = q_f/q_f^0$, for several values of the degree of saturation (C_A^0). For each curve $\phi_0 = 100$. Other parameter values appear in Table 1.

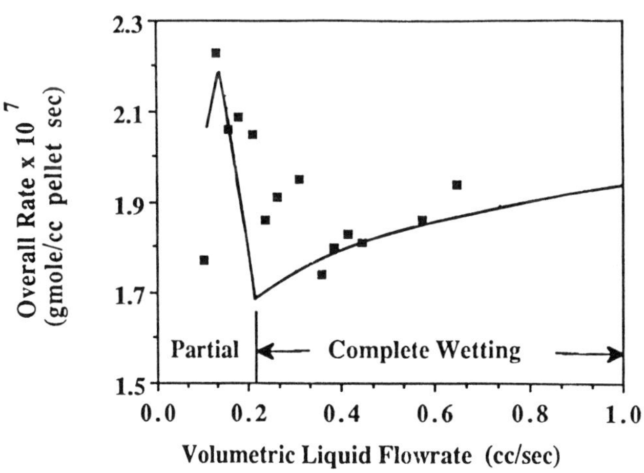

Figure 4. Model simulation (line) of the data (points) of Satterfield et al. (1969) for α-methylstyrene hydrogenation on a string of fourteen palladium on alumina spherical pellets. Parameter values used are provided in Table 2.

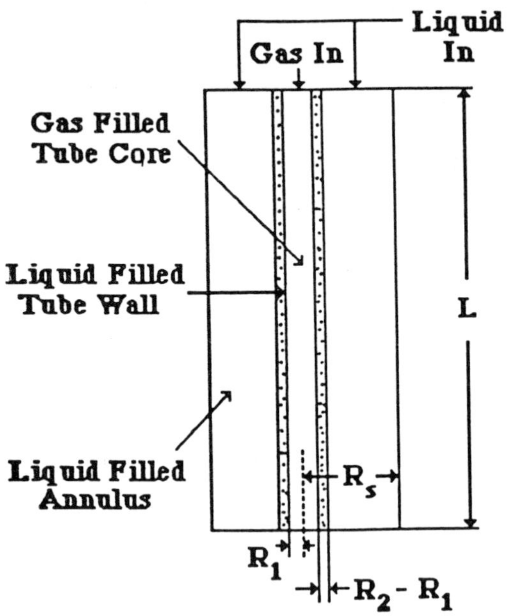

Figure 5. Schematic of the catalytically impregnated tube contacted cocurrently by the gas in tube core and the liquid on the shell side.

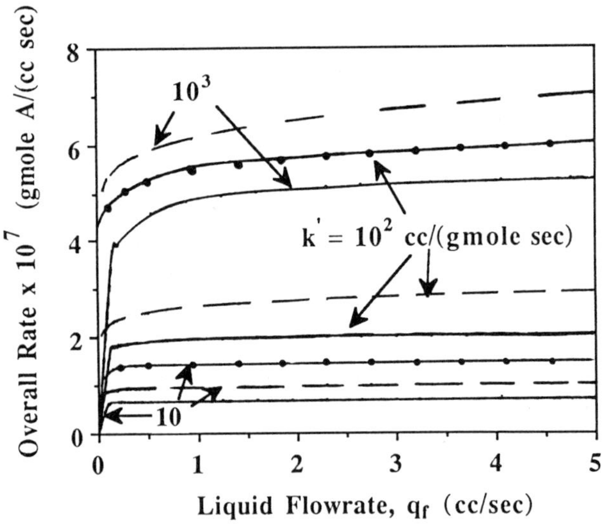

Figure 7. A comparison of the overall rate as a function of the liquid flowrate for the pellet string and two different catalytic tubes contacted by cocurrent gas-liquid flow. The catalyst volume and overall length are fixed and the catalytic activity (k′) assumes three different values. The feed liquid is saturated with A and contains a large supply of B (c_{Bf} = 10^{-3} gmole/cc). Catalyst dimensions and other parameter values are provided in Table 4.

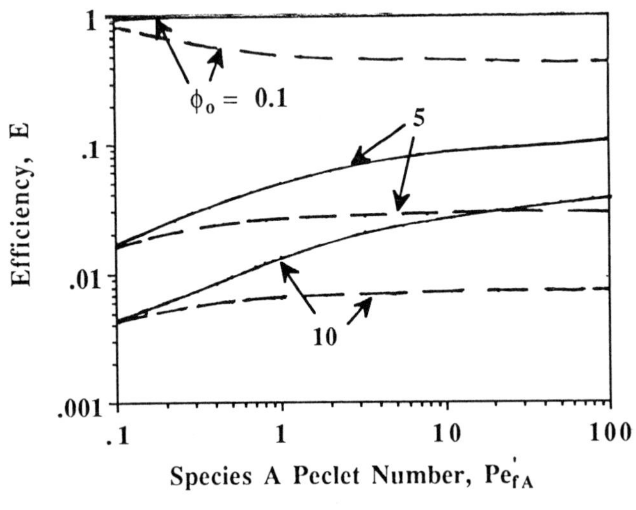

Figure 6. Efficiency of the cocurrent flow catalytic tube as a function of the liquid flowrate for several values of the Thiele modulus (ϕ_0). The solid curves are for the feed liquid containing its equilibrium allotment of volatile reactant (C_A^0 = 1). For the dashed curves C_A^0 = 0.01. Other parameter values appear in Table 3.

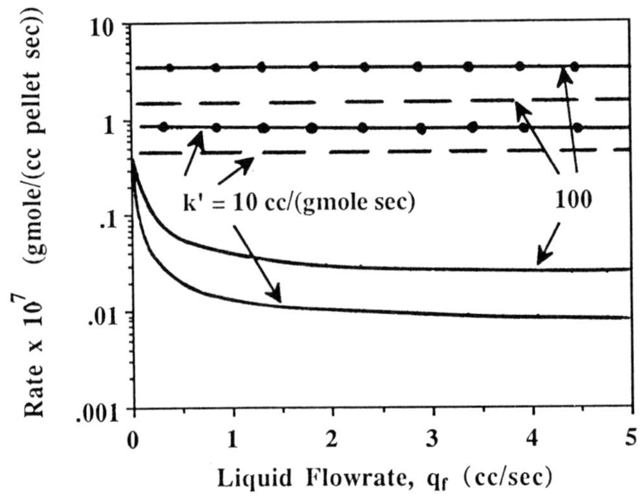

Figure 8. A comparison of the overall rate as a function of the liquid flowrate for the pellet string and the two different catalytic tube as in Figure 7. The catalyst volume and overall length are fixed and the catalytic activity (k′) assumes three different values. The feed liquid contains a large supply of B(c_{Bf} = 10^{-3} gmole/cc) and only a 1% allotment of A. Catalyst dimensions and other parameter values are provided in Table 4

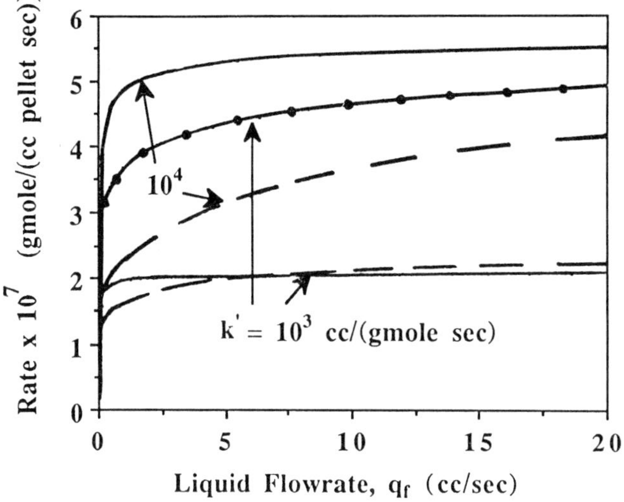

Figure 9. A comparison of the overall rate as a function of the liquid flowrate for the pellet string and the two cocurrent flow catalytic tube reactors as in Figures 7 and 8. The catalyst volume and overall length are fixed. The catalytic activity (k′) assumes two different values. The feed liquid does not contain a large supply of B(c_{Bf} = 10^{-4} gmole/cc) but is saturated with A. Catalyst dimensions and other parameter values are provided in Table 4.

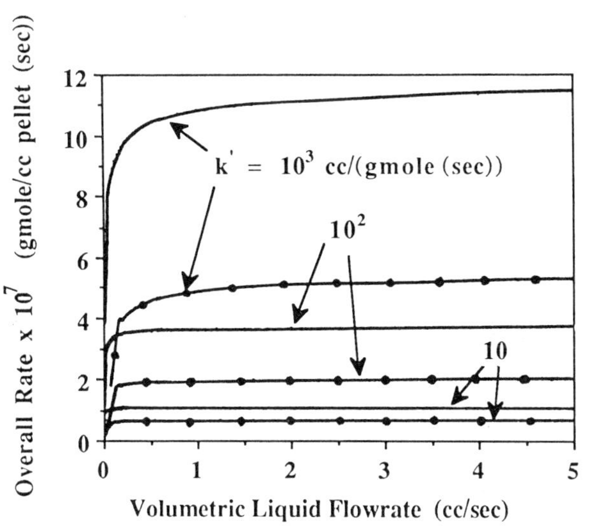

Figure 10. A comparison of the overall rate as a function of the liquid flowrate for the pellet string and a catalytic tube contacted by a well stirred liquid. The catalyst volume and overall length are fixed and the catalytic activity (k′) assumes two different values. The feed liquid contains a large supply of B(c_{Bf} = 10^{-3} gmole/cc) and is saturated with A. Catalyst dimensions and other parameter values are provided in Table 4.

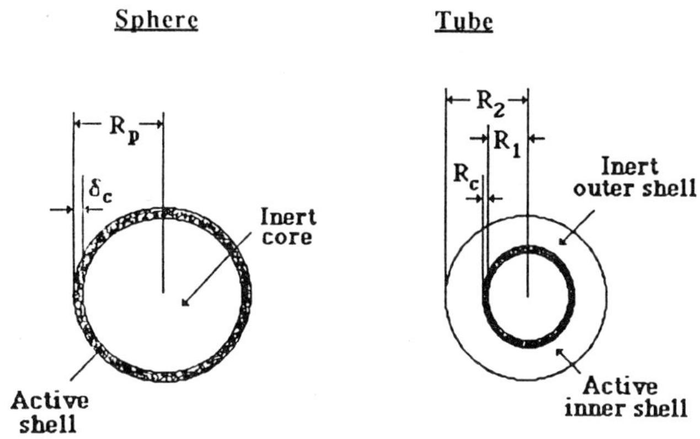

Figure 11. Schematic of the surface shell impregnated sphere and tube.

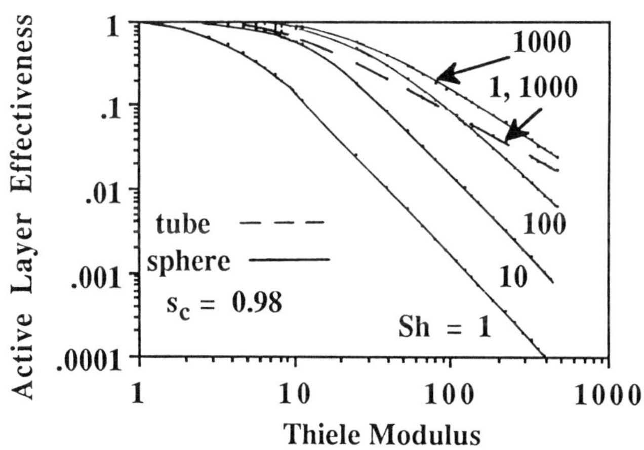

Figure 12. Comparison of the active layer effectiveness as a function of the Thiele modulus based on the overall diffusion length for the surface shell impregnated sphere and tube.

Table 1. Set of parameter values used in simulation of string of pellets (Figures 2 and 3).

$a_A = 3 \times 10^{-4}$	$\lambda_s = 2$
$a_I = 5 \times 10^{-6}$	$\lambda_p = 10$
$b_A = 10^8$	$P_G = 10^3$
$b_I = 2.5 \times 10^7$	$\beta_{oA} = 0.02$
$\Lambda = 5 \times 10^3$	$Bi_{gsA}{}^o = 10^4$
$\alpha = 3\pi$	$\nu = 1$
$\Delta = 0.5$	$P_f\,(Q = 1) = 10^5$
$C_A{}^o = 10^{-2}, 0.1, 0.5, 1$	$\phi_o = 0.1 - 200$
$C_f{}^o = 25 \times 10^3, 250, 50, 25$	

Table 2. Values used for the simulation of the Satterfield et al. (1969) data (Figure 4). Variables not provided by the investigators which we have estimated are denoted with an asterisk.

$D_{Ae} = 1.096 \times 10^{-5}\ cm^2/sec$	$H = 10.65$
$^*D_{Be} = 5 \times 10^{-6}\ cm^2/sec$	$q_G\ (z = 0) = 1.56\ cm^3/sec$
$k' = 2.24 \times 10^3\ cm^3/$ gmole B sec	$\nu = 1$
$^*\mu_G = 10^{-4}$ g/cm sec	$R_p = 0.4125$ cm
$R_{p,core} = 0.1625$ cm	$R_{p,act} = 0.25$ cm
$c_{Ae} = 3.54 \times 10^{-6}$ gmole A/cm^3	$^*c_{Bf}(z = 0) = 7.5 \times 10^{-3}$ gmole B/cm^3
$R_s = 2.67$ cm	$^*\mu_f = 0.135 \times 10^{-2}$ g/cm sec
$L = 14(2R_p) = 11.55$	$\sigma = 30$ dyne/cm
$x_{AG}\ (z=0) = 1$	$\rho_f = 0.908\ g/cm^3$
$M_A = 2$	$^*w = 0.149$
$T = 323.15$ K	$^*\theta = \pi/2$
$P_T = 1$ atm	$\varepsilon_p = 0.498$
$^*k^o_{gsA} = \kappa k_{gl}$; $\kappa > 1$	$\varepsilon_p/\tau = 0.066$

Table 3. Parameter values used in simulation of cocurrent flow catalytic tube (Figure 6).

$a_A = 10^{-2}$	$\Delta_A = 0.1$
$a_I = 1.5 \times 10^{-4}$	$\Delta_B = 0.1$
$b_A = 10^6$	$Pe_G = 200$
$b_I = 2.5 \times 10^5$	$\hat{\xi}_s = 10$
Pe'_{fA}: $0.1 - 10^3$	$\hat{\xi}_1 = 3$
$Pe'_{fB} = 2\ Pe'_{fA}$	$\hat{\xi}_2 = 5$
$\phi_o = 0.1 - 10$	$m_o = 0.5$
$C_A{}^o = 10^{-2} - 1$	

Table 4. Set of values used in the performance comparison (Figures 7-10).

Values common to all models:

$\nu = 1$	$x_{AG}(z = 0) = 0.5$	$D_A = 10^{-4}\ cm^2/sec$
$D_B = 5 \times 10^{-5}\ cm^2/sec$	$D_{Ae} = 10^{-5}\ cm^2/sec$	$D_{eB} = 5 \times 10^{-6}\ cm^2/sec$
$\mu_G = 10^{-4}$ g/cm sec	$H = 1$	$\varepsilon_p = 0.5\ cm^3$ void/cm^3 pellet
$p_T = 1$ atm	$L = 25$ cm	$T = 323.15$ K
$C^o_A = 0.01 - 1$	$q_G(z=0) = 1\ cm^3/sec$	$k' = 1 - 10^4\ cm^3$/gmole sec
$M_A = 2$ g/gmole	$M_I = 28$ g/gmole	

Additional values for the string of pellets:

$R_p = 0.25$ cm	$c_{Bf}(z = 0) = 10^{-3} - 10^{-4}$ gmole/cm^3
$\mu_f = 5 \times 10^{-2}$ g/ cm sec	$q_f = 0 - 10\ cm^3/sec$
$\rho_f = 0.9\ g/cm^3$	$\sigma = 30$ dyne/cm

Additional values for the cocurrent flow catalytic tube:

$R_2 - R_1 = 0.1, 0.05$ cm	$c_{Bf}(z = 0) = 10^{-3} - 10^{-4}$ gmole/cm^3
$R_1 = 0.1583, 0.3917$ cm	$q_f = 0 - 20\ cm^3$ /sec

Additional values for the well stirred liquid catalytic tube:

$R_2 - R_1 = 0.1, 0.05$ cm	$c_{Bf}(z = 0) = 10^{-3} - 10^{-4}$ gmole/cm^3
$R_1 = 0.1583, 0.3917$ cm	$q_f = 0 - 20\ cm^3$ /sec
$Sh_A = 100\ q_f^{0.5}$	$Sh_B = (D_A/D_B)^{0.5}\ Sh_A$

INORGANIC MEMBRANE REACTORS —A REVIEW

H.P. Hsieh ■ Alcoa Laboratories, Alcoa Center, PA 15668

The use of membranes as both separators and reactors in a single unit operation for some catalytic reactions is capable of increasing yield and, in some cases, selectivity due to equilibrium displacement by the continuous removal of one or more products or suppression of undesirable side reactions by controlled addition of the reactants. With their inherent thermal, structural, and chemical stabilities, inorganic membranes are particularly attractive for many high temperature gas phase reactions or those reactions in corrosive environments.

The preparation methods and structural characeristics of nonporous membranes made from palladium and its alloys, nickel, silver, and zirconia and porous membranes made from porous glass, alumina, zirconia, and other oxides are briefly discussed. Current and potential applications, in both liquid and gas phases, of these membranes are identified.

The technology of inorganic membrane reactors is reviewed in the areas of thermal and mechanical stabilities of membranes, selective permeabilities, catalyst impregnation, membrane/reaction considerations, reactor configuration, and reaction coupling. Existing literature information, mostly derived from research institutions outside the U.S., has demonstrated the feasibility of the concept of using inorganic membranes in catalytic reactions such as hydrogenation (or hydrodealkylation) and dehydrogenation of many gaseous compounds, decomposition of hydrogen sulfide, decomposition of water, steam reforming of methane, and conversion of sucrose to glucose.

The technology of organic membranes, accelerated by the needs of desalination in the 1970s, has advanced from liquid separation to gas separation applications and become an acceptable and, in many cases, economically viable unit operation. In contrast, inorganic membranes, although having been studied since the late 1950s, have just begun to gain recognition as a separation tool with the introduction of industrial-scale, structurally stable and consistent-quality porous ceramic membranes such as alumina membranes. These membranes are characterized by separative ceramic films of controlled porosity and morphology. The increasingly studied technology of inorganic membranes has been reviewed recently (1, 2, 3).

Many heterogeneous catalytic reactions can not reach high conversions as a result of the limits imposed by the reaction equilibrium. For example, the conversion of the dehydrogenation reaction of cyclohexane catalyzed with Pt/Al_2O_3 at 215°C in a conventional packed bed is limited to 33%. If, however, the product hydrogen (but essentially no cyclohexane or benzene) is continuously removed from the reaction mixture through a glass membrane, the equilibrium is displaced toward the product side and the conversion has been shown to increase to 80% (4). Thus, the combination of reaction and membrane separation (membrane reactor) is capable of increasing the reaction yield beyond what the reaction equilibrium allows. It is noted that the increase in conversion at a given temperature can mean a lower operating temperature required to achieve a given conversion.

In some cases, as will be discussed later, membrane reactors can also be used to improve the product selectivity. Some general advantages of using membrane reactors over the conventional packed bed reactors are summarized in Table 1. The potentials of membrane reactors for improving yield and selectivity and for integrating two unit operations, reaction and separation, into one are being recognized industrially (5).

Most of the catalytic reactions, particularly those in petroleum and petrochemical industries, are carried out at elevated temperatures exceeding the maximum operable temperatures (generally approximately 150°C) for organic membranes. In addition, some catalytic reactions create corrosive environments due to combinations of reactants, intermediates, products and operating conditions. In view of their inherent thermal and chemical stabilities, many inorganic membranes are good candidates for use in those reactions.

The purpose of this paper is to review the status of inorganic membrane technology and how inorganic membranes can improve conversion and/or selectivity of some heterogeneous catalytic reactions.

COMMERCIAL AND DEVELOPMENTAL INORGANIC MEMBRANES

Commercial Membranes

Inorganic membranes can be conveniently divided into two categories: dense and porous membranes. Dense inorganic membranes, with the exception of palladium membranes utilized in the Soviet Union (6), have not been widely used in the industry. Currently commercial porous inorganic membranes include alumina, glass, silver, stainless steel and zirconia membranes as shown in Table 2. They come in three general geometries: disk or plate, tube and multichannel monolith. The membrane pore diameter ranges from approximately 40 Å to 10 µm, covering both ultra- and microfiltration applications.

Developmental Membranes

Although not yet commercialized, many inorganic materials have been considered as candidates for making membranes. Table 3 provides some examples of the developmental efforts.

Some ceramic materials such as titania, mullite, cordierite, mica, silicon carbide and silicon nitride have been studied. Others include tin oxide, smectite, etc. They represent a wide range of pore size and morphology, structural stability and affinity for water or organic solvents.

Porous metals and alloys have also been suggested: transition metals and their alloys, and amorphous metal alloys. Zeolite, molecular sieve carbon and polyphosphazenes are new additions to the list of developmental membrane materials.

Preparative Methods

Some dense metals and metal oxides are known to be selective only to certain gases such as hydrogen or oxygen. Preparation of these materials as membranes involves pressing, high-temperature sintering and/or melt processing. Their methods of preparation are well-established technology and will not be repeated here. Only those methods for making porous inorganic membranes are discussed.

Most ceramic membranes are made by the slipcasting method. It involves first preparing a stable dispersion of fine particles called slip and then depositing a layer of particles on a porous support by the capillary pressure drop created by the contact of the slip with a dry microporous support. The capillary pressure drop forces the dispersion medium (e.g., water) to flow into the support pores, thus concentrating the particles to form a membrane layer. To prepare nanometer size particles required for making membranes with pore diameters of a few nanometers, the sol-gel process can be employed (23, 24, 25). After the membrane layer is formed, careful drying and calcination are required to arrive at the desired membrane pore size without suffering from the problems of cracks or pinholes.

A porous glass membrane can be prepared by the phase separation/leaching method (26). Heating is applied to the glass composition to effect the separation of two homogeneously distributed phases: a SiO_2-rich phase and a Na_2O-B_2O_3-rich phase. The latter phase can be water or acid leached and the glass becomes a microporous structure. Some porous metal membranes have been made by a similar approach (22).

Anodic oxidation can be used to convert high-purity, thin metal foils (e.g., aluminum or zirconium) into porous oxide membranes having approximately straight pores (27, 28, 29, 30, 31). One side of the metal foil is oxidized anodically in an acid electrolyte such as phosphoric, sulfuric or oxalic acid. This is followed by removing the unaffected part of the metal from the porous film by an etchant (e.g., hydrochloric acid), a dissolving compound (e.g., bromine/methanol), mercuric chloride, or an electrochemical process to eliminate closed pores (31). The resultant membrane pores can be symmetric or asymmetric across the membrane thickness. Presently, this method is limited to making small piece membranes suitable for laboratory bench filtration.

The track-etch method has been used to make porous mica membranes. A radioactive radiation source is applied to fire particles through mica. The track left behind can be easily etched with an etchant such as concentrated HF to yield a membrane characterized by well-controlled straight pores having uniform shape and size (11, 12).

Some inorganic membranes can be made by controlled pyrolysis of organic materials which provide the basic framework for the final porous membranes. For example, silica and carbon molecular sieve membranes have been made by carefully pyrolyzing, respectively, silicone rubbers and certain thermoset polymers (32, 18).

The thin film deposition technology has been applied to making porous membranes. Teijin (21) demonstrated that transition metals or their alloys can be deposited on porous supports by using physical or chemical deposition techniques such as sputtering, ion plating, metal plating or vapor deposition.

Inorganic polymers present vast opportunities for preparing high-temperature inorganic membranes with various pore sizes. McCaffrey (20) studied the synthesis of polyphosphazenes and characterized them for gas separations.

Structural Characteristics

Some porous glass and metal membranes have a uniform structure across the thickness of the membranes. This uniform thick structure, although contributing to the physical integrity of the membranes, is not desirable from the flux viewpoint. An ideal membrane should be as thin as possible, to reduce the flow resistance, and, on the other hand, sufficiently thick and strong to withstand the applied transmembrane pressure difference, fluid dynamic shear and any normal handling and cleaning. Therefore, as a workable compromise, most of the porous inorganic membrane elements have an asymmetric structure: a thin separative membrane layer and one or more comparatively thick support layers.

One of the major reasons for the increasing awareness and interest in inorganic membranes in recent years is the breakthrough of 5-20 μm thick, consistent-quality, pinhole-free ceramic membrane layers as exemplified in Figure 1. The use of multiple support layers is also attributed to the flux consideration. If one were to deposit nanometer size particles (for the nanometer pore size membrane layers) directly on a porous support having a pore diameter of, say, 10 μm, the result would be appreciable penetration of the ultrafine particles into the support causing blockage of many pores. Thus, the choice of a graded structure of the membrane elements represents an approach to maximizing the permeate flux subject to the constraints of mechanical strength of the membranes. The graded structure is reflected in a typical pore size distribution shown in Figure 2 for an alumina membrane (2). A very narrow pore size distribution in each layer is evident.

In contrast to tortuous flow paths through most of the porous inorganic membranes, those membranes made by the anodic oxidation or the track-etch method exhibit more or less straight pores. They have been used as model systems in fundamental transport studies. Presently, only anodic aluminum oxide membranes are commercially available for small-scale laboratory applications.

Separation Applications

Some dense metal membranes such as those made of palladium and its alloys have been used industrially in the Soviet Union for separating hydrogen from other gaseous or liquid compounds (6). Dense zirconia membranes have been used on a limited scale in oxygen sensors.

So far most of the industrial applications of inorganic membranes lie in the area of liquid phase separations and largely involve porous ceramic membranes: alumina and zirconia. Both porous alumina and zirconia membranes have also been used earlier in separating uranium isotopes in the nuclear industry. With that exception, porous inorganic membranes have not yet seen commercial uses in gas separations although there have been many studies (2). Some examples of studying the use of inorganic membranes for gas separations are given in Table 4.

Proven liquid-phase applications of the various inorganic membranes have been discussed elsewhere (2, 3) and a cross section of those applications is given in Table 5. Commercial porous alumina, zirconia, stainless steel, silver, and glass membranes offer a water permeability of approximately 3,000 to 5,000 L/h-m^2-bar for a pore diameter of 0.2 to 1 μm in the microfiltration applications. The typical water permeability for ultrafiltration membranes with pore diameters of 20-500 Å ranges from 1 to 250 L/h-m^2-bar. Generally the cross-flow velocity used is 1 to 7 m/s.

The ability to steam sterilize and autoclave many inorganic membranes to meet cleaning requirements is a significant advantage in many applications. Recently introduced ceramic membranes, particularly alumina membranes, also can be backflushed with short pulses of permeate on a periodic basis to reduce the typical problem of flux deterioration due to gel formation or concentration polarization.

CURRENT STATUS OF INORGANIC MEMBRANE REACTORS

The technology of inorganic membrane reactors is still at its infancy stage, perhaps with the exception of some dense palladium-based membrane reactors used to some extent in industrial production of chemicals and pharmaceuticals in the Soviet Union. However, there has been an increasing volume of research exploring the use of inorganic membranes as catalytic reactors, particularly in Japan and the Soviet Union.

Dense Inorganic Membrane Reactors

Most of the extensive studies of dense inorganic membrane reactors are related to gas-phase, high-temperature catalytic reactions - hydrogenation and dehydrogenation of various hydrocarbons. And the majority of them use palladium or its binary or ternary alloys as catalytic membranes. Some typical examples are given in Table 6.

The popular uses of palladium-based membranes are the results of the 1866 discovery by Thomas Graham (44) that metallic palladium absorbs an unusually large amount of hydrogen. Hydrogen permeates through Pd-based membranes in the form of highly active atomic hydrogen which can react with other compounds adsorbed on the catalyst surface. In many cases, palladium alloys are preferred over pure palladium for two major reasons. First, some selective palladium alloys offer greater catalytic activity and hydrogen permeability than pure palladium (45). Second, pure palladium can become brittle after repeated cycles of hydrogen absorption and desorption (46). The metals most used as the secondary or ternary components in the palladium alloys are Ru, Rh, Ni, Cu and Ag.

Gas-Phase Reactions. In dehydrogenation reactions where one of the products, hydrogen, is continuously removed from the reaction zone, the conversion can be significantly increased. For example, continuously removing hydrogen from the dehydrogenation product of isopentenes in a Pd-Ni alloy membrane reactor can improve the yield of isoprene to 21% compared to 14% when the reaction takes place in a fixed-bed reactor (41). Itoh (36) studied dehydrogenation of cyclohexane in a Pd membrane reactor containing a packed bed of Pt/Al_2O_3 catalyst. The equilibrium conversion was only 18.7%, but the removal of hydrogen from the reaction mixture using the membrane increased the conversion to as high as 99.5%.

The removal of a reaction component from a dense membrane reactor can also enhance the selectivity. In the hydrodealkylation of 1,4-and 1,5-dimethylnaphthalene, the undesirable byproduct ß-methylnaphthalene can be suppressed by removing hydrogen through a Pd-Ni alloy membrane. The resultant products are the desirable α-methylnaphthalene and naphthalene (41).

For hydrogenation reactions where hydrogen is added to the reaction through a hydrogen selective membrane, selectivity can be greatly improved by controlling any undesirable side reactions. For example, when propylene and oxygen react in a conventional fixed bed reactor at 600°C with a bismuth oxide-based catalyst, the selectivity to C_6 dimers after 24 hours is only 32-39% (with the rest being CO, CO_2, CH_4, etc.) as a result of prolonged contact of oxygen with the hydrocarbons to form undesirable carbon oxides. If, however, a bismuth oxide-based membrane permeable only to oxygen is used to separate the reaction zone from the oxygen, the reaction selectivity at the same temperature and under otherwise similar conditions is increased to 77% (47). Thus, controlled addition of an overly active reactant to a reactor through the use of a permselective membrane can significantly improve the reaction selectivity.

But in some cases, the conversion rate is also increased by adding reactants to opposite sides of a membrane which is selective to only one reactant. For example, the calculated equilibrium yield of cyclohexane from the hydrogenation of benzene is 0.1% at 390°C, but controlled addition of hydrogen through a Pd membrane can increase the yield to 4% (41). The hydrogenation reaction of 1,3-pentadiene using a dense membrane made of a palladium alloy with 9.8% ruthenium proceeds approximately 2.4 times faster than that using a premixed gas mixture of hydrogen and 1,3-pentadiene (42). The main product with the use of a catalytic membrane is 1-pentene while that without the membrane is cis-2-pentene. Nagamoto and Inoue (34) conducted hydrogenation experiments of 1,3-butadiene and found that hydrogen permeation through a palladium membrane increased the reaction rate to a value more than ten times that when hydrogen was premixed with butadiene. The remarkable increase in the conversion rate was attributed to the higher surface coverage of hydrogen on the catalyst as a result of the high reactivity of the dissolved hydrogen in the palladium membrane compared to the slower hydrogen adsorption in the case of premixed hydrogen and butadiene gases.

Dense membranes made of other metals such as silver have been used in oxidation of ethylene (48), oxidation of ethanol and propene (49) and oxidation of ammonia (50). Silver has been known to be selectively permeable to oxygen.

Dense zirconia, especially calcia-stabilized, membranes are also preferentially permeable to oxygen. Therefore, stabilized zirconia membranes have been used as catalytic membranes for a number of reactions. They have been shown to selectively transfer oxygen and increase the conversion in the thermal decomposition of water to produce hydrogen (51) and in the steam reforming and shift reactions to form hydrogen (52). Nigara (53) used 10% calcia-stabilized dense zirconia membrane reactors for the direct thermal decomposition of carbon dioxide at temperatures up to 1680°C to produce carbon monoxide. The use of the membrane reactors increased the dissociation rate of the carbon dioxide from 1.2% as the limit imposed by the reaction equilibrium to 21.5% following continuous removal of oxygen from the reaction mixture.

Liquid-Phase Reactions. Palladium and its alloy membranes have also been used in liquid phase hydrogenation reactions such as transformations of aliphatic alcohols. Some examples include hydrogenation reactions of 2-butyndiol-1,4 and of dehydrolinalool (42). In addition to liquid phase hydrogenation, inorganic membranes have also been used in or as membrane reactors in liquid phase electrochemical reactions and bioreactions. Guther and Vielstich (54) studied electrocatalysis of CHO-compounds. On one side of a palladium membrane they deposited platinum as the catalyst for the dehydrogenation of methanol, formic acid or formaldehyde dissolved in an electrolyte. Hydrogenation of sodium p-nitrophenolate to produce sodium p-aminophenolate was investigated in a polarized nickel membrane reactor as a function of the reaction conditions including the potential and temperature (55).

Porous Inorganic Membrane Reactors

Studies of porous inorganic membranes in or as membrane reactors have started only recently with most of them conducted in Japan. Representative studies are summarized in Table 7 which clearly shows the feasibility of employing various porous inorganic membranes for increasing the conversion of some reactions. Under the optimal conditions of dehydrogenation of cyclohexane, for example, the membrane-enhanced conversion can be 250% of the equilibrium-limited conversion (36).

The catalytic reactions which have been investigated using porous membranes include decomposition of hydrogen sulfide (57, 58, 61), decomposition of hydrogen iodide (64, 65), dehydrogenation of cyclohexane (4, 36), (oxidative) dehydrogenation of methanol (60), dehydration of isopropanol (59) and decomposition of RuO_4 (63). The membrane materials tested range from glass, alumina to mixtures of Fe_2O_3, molecular sieves and clay.

Based on available data on temperature and membrane pore diameter which is approximately equal to or larger than 40 Å, the predominant separation mechanism should be surface diffusion and/or Knudsen diffusion, particularly the latter. The separation factor due to Knudsen diffusion is inversely proportional to the square root of the molecular weights. When the permeate is hydrogen, as in most of the studied cases, the resultant separation factor can be great enough to make the approach a technically viable one. The separation factor under the Knudsen diffusion regime increases markedly with rising temperatures. On the other hand, when the reaction gases do not differ much in molecular weight, particularly at low temperatures, Knudsen diffusion is not effective for gas separation.

Catalysts such as Pt, Pt/Al_2O_3, Al_2O_3, MoS_2, WS_2, Fe_2O_3 and enzyme have been introduced into the membrane reactors used. The catalysts can be incorporated in a packed bed retained by the membrane, impregnated or immobilized on the membrane surface or in the membrane pores or as the integral part of the membrane matrix. It is interesting to note that in the case of impregnation, the point of introduction can be the high-pressure (feed) or the low-pressure (permeate) side of the membrane and the results are rather different (59).

The fluid dynamics, as expected, can have strong impacts on the reaction conversion. For example, it has been shown (4, 36, 62) that, for dehydrogenation of cyclohexane, as the flow rate of the feed gas through the membrane reactor is lower or that of the purge (and permeate) gas outside the membrane reactor is higher, the yield becomes higher. This can be predicted by a model (4) and can be attributed to the constant diffusion driving force across the membrane and the elimination of any back diffusion from the permeate side to the retentate side, as a result of sweeping action by the flowing purge gas. The same phenomenon occurs in an alumina membrane reactor with transition-alumina as the catalyst for the dehydrogenation reaction of methanol (60). Itoh et al. (64, 65) studied the catalytic decomposition of hydrogen iodide and dehydrogenation of cyclohexane in a porous glass membrane by computer simulation. They demonstrated that: (1) faster removal of the product gas hydrogen results in a higher conversion; and (2) the optimum membrane thickness for the maximum conversion is determined by the relative rates of permeation and reaction.

ATTRIBUTES OF EFFECTIVE INORGANIC MEMBRANE REACTORS

To make the unique combination of inorganic membranes and reactors effective, some requirements related to membrane and catalytic properties can be summarized as follows.

Selective Permeability

Essential to the concept of membrane reactors is the ability of the membrane to be selectively permeable to one or more components present in the reaction product stream. A good separation factor of some more permeable product(s) from the remaining less permeable components will significantly displace the reaction equilibrium toward the product side, thus enhancing the reaction conversion. The operating condition of the reactor should take advantage of some possible separation mechanisms of the membrane, such as size exclusion, Knudsen and surface diffusion and capillary condensation (66) to be discussed later. In addition to separation factor, the permeate flux is also very important which is directly related to throughput.

Thermal and Structural Stabilities

The inorganic membrane used should be able to withstand the maximum temperature in the course of the reaction. For example, some porous glass membranes become unstable structurally above approximately 800°C (57). For some dense membranes made of metals such as palladium, repeated cycles of sorption and desorption of hydrogen in its normal usage are known to make the metal brittle (46). Swelling of dense metallic membranes can also occur.

Catalytic Activity

The required catalytic activity for the reaction must be introduced internal or external to the membrane. When the catalyst(s) is an integral part of or impregnated in the membrane, the membrane can be a catalyst or catalyst carrier. The term catalytic membrane used here refers to a membrane which by itself is catalytic to the reaction under consideration without the use of additional catalyst(s). When a candidate material such as alumina, silica, molecular sieves, or other metal (Pd, Ni, Pt, etc.) is considered for a catalytic membrane, its catalytic activity should be compatible to the particular reaction(s) involved. The catalytic activity of a catalytic membrane can be influenced by the particular phase of the membrane (42) or the fabrication history (6).

There are several methods for preparing catalysts: impregnation, precipitation, adsorption and dry mixing. The catalyst, in the case of a catalytic membrane, is incorporated in the porous separator (membrane) and uses the membrane surface or pore surface as the catalyst carrier or catalyst. The porous nature of the membrane makes impregnation and adsorption preferred choices.

Reaction Rate

Typically the residence time of the feed stream in a membrane reactor is short (less than a few minutes). Therefore, those reactions with relatively fast kinetics are more likely to utilize the concept of membrane reactors.

MEMBRANE/REACTION CONSIDERATIONS

Membrane Permeability and Permselectivity of Reactants versus Products

Proper choice of the membrane can be greatly affected by the reaction involved. An important factor is the relative permeabilities of the reactants and products through the membrane. Consider, for example, the following reversible reaction:

$$A + B \rightleftharpoons P + Q.$$

When only one product (e.g., P) is permeable, the conversion becomes higher as the equilibrium is displaced toward the product side. If some side reactions are present, the removal of the permeable product will tend to suppress the side reactions, thus achieving selectivity control. If, on the other hand, only one reactant is permeable, the removal of the reactant shifts the equilibrium toward the reactant side and results in a lower conversion. It could also create unbalanced stoichiometry conditions.

If only both products are permeable through the membrane, the equilibrium will drive the conversion higher. It also results in the desired reactants/products separation. When only both reactants permeate through the membrane, a lower conversion is expected. If the membrane is porous and the reactants react on the pore surface, pore blockage may occur due to the larger size of the product molecules.

Membrane Operating Conditions and Reaction

Since very often a pressure difference across the membrane is used as the driving force for separating reaction components, the operating pressure or the partial pressure is likely to have some effects on the reaction conversion and selectivity. In a study of the hydrogenation of 1,3-butadiene, Nagamoto and Inoue (34) found that, for a given membrane, the butadiene pressure has significant impact on conversion and selectivity. At low pressures (e.g., <20 kPa) of butadiene, the reaction rate is lower and the selectivity for butene formation is also lower. At a given butadiene pressure, the hydrogenation rate, but not the product composition, strongly depends on the upstream hydrogen pressure.

In addition to pressure, the contact time between the reactants and the membrane catalyst can also affect the reaction path and selectivity (60).

MEMBRANE REACTOR CONFIGURATION AND GEOMETRY

An inorganic membrane can be configured in many possible ways with the catalytic reactor. First, the membrane is a separate physical entity either upstream or downstream of the reactor. The membrane can also be a part or the whole of the reactor. For example, the membrane can be part of the container for the catalyst or catalyst carrier or it can itself be the catalyst or catalyst carrier. When the membrane is also the catalyst, it is referred to as a catalytic membrane.

The common tubular and disk shapes have low packing densities (low filtration area per unit volume) but are easy to maintain. For example, tubular ceramic membrane bioreactors are described in a recent U.S. patent (67). In essence, a two-layered tubular membrane element was used in which the inner layer had larger pores than the outer layer. Enzymes were immobilized inside the inner layer with a fixing agent such as silanes or olefins. The reaction solution is pumped into the hollow central core of the tube and, under pressure, flows into the inner layer where the solution reacts with the enzyme. The products, but not the enzyme, pass through the selective outer layer and exit the membrane composite. Thus, the enzyme is retained in the inner layer. The packing density can be improved by the use of multichannel monoliths.

The membrane reactor can also be a flat membrane as described in a recent patent (59) where the flow-through instead of the cross-flow pattern is used. The catalyst can be deposited within the pores, preferably in the ends of the pores close to the downstream surface of the membrane.

Gryaznov (6) has shown two other higher packing density geometries of large-scale palladium-based membrane reactors. In one geometry, a membrane is fabricated into a spiral plate, forming a double spiral with two compartments on the opposite sides of the membrane. The edges of the plate are built into the walls of the reactor shell. In another geometry, a membrane is made of seamless thin-walled tubes to form double flat spiral partitions. The spirals are positioned one on top of the other so that each succeeding spiral is in a mirror symmetry with the adjacent one. This enables the maximum number of spirals to be packed into a given reactor volume.

Due to the inherent rigidity of inorganic membranes, their presently available geometries as reactors are not as diverse as those of organic polymeric membranes. In addition to disks, tubes and monoliths, other higher packing-density geometries (e.g., hollow fiber (68)) of inorganic membranes have been attempted but not commercialized, probably due to strength considerations.

A rather intriguing extension of the concept of membrane reactors is the so-called reaction coupling (69). When a product selectively permeating through a membrane reactor enters and is used as a reactant in a second membrane reactor situated on the low pressure side of the first membrane, as shown in Figure 3, the two reactors are coupled to handle two reactions in one unit operation. Thus the membrane serves as: (1) a chemical separator; (2) a partition to prevent the reaction contents in both membrane reactors from intermixing; and (3) a heat transfer medium for the two coupled reactions when one is endothermic and the other exothermic. The concept has been demonstrated with dehydrogenation of olefins and hydrogenation of aromatic hydrocarbons (6). The coupling can also be done with dehydrogenation in one reactor and combustion in the other, both reactors sharing a common membrane (41).

SOME KEY TECHNOLOGICAL AREAS

As the interests for inorganic membrane reactors are intensifying, some basic and practical technological issues need to be addressed before large scale implementations become a reality. Three such issues are discussed in the following.

Membrane Pores

As mentioned earlier, some effective separation mechanisms through a porous inorganic membrane include molecular size exclusion, Knudsen diffusion, surface adsorption and diffusion, and capillary condensation.

For a membrane to be selective based on the molecule size difference, the membrane pores need to be narrowly distributed and, in many cases, rather small when both the permeate and the retentate have comparable molecular sizes. Presently commercial porous inorganic membranes are limited to a minimum pore diameter of approximately 40 Å. Unmodified pores in this size range are able to utilize Knudsen diffusion for separation, but are likely to be out of the effective operating domains of other separation mechanisms in many situations. Attempts to make smaller pore size membranes are evident. Zeolite or molecular sieve membranes are some examples (14, 15, 16, 18). Pore sizes can also be effectively reduced by controlled modifications to the membranes, for example, by precipitation in the pores (70). It is noted that when a catalyst is impregnated inside the membrane pores, the pore size may become appreciably smaller.

For surface adsorption and diffusion to be effective, the pore diameter generally needs to be small (10-100 Å) or the specific surface area of the pores needs to be large. An alternative is to modify the membrane pore surfaces to render them more affinitive to the permeable molecules. There has been an increasing amount of research on modifications of inorganic membranes recently (71, 72, 73, 74, 75).

When the operating temperature is near 0°C and/or the pressure is high, capillary condensation can occur with the proper combination of gases. At high temperatures which are common in many gas phase reactions, this phenomenon is not likely to be significant.

Catalysts

Zaspalis (60) found that bimolecular reactions, at least in the case of dehydrogenation of methanol, did not seem to be favored within the membrane pores. While this may not be the case with other reactions, it points out the importance of how the catalyst is introduced to the membrane. The location at which the catalyst is impregnated (e.g., at the entrance or the exit of pores) has been found to influence the effectiveness of the membrane reactor (59).

The membrane material itself is often catalytically active to some reactions. It has been mentioned earlier that palladium is catalytically active to many hydrogenation and dehydrogenation reactions. Transition-alumina has been proven to be an effective membrane catalyst for the (oxidative) dehydrogenation of methanol (60). Alumina is also known to be a catalyst for many reactions such as alkylation of phenol, dehydration of formic acid, isomerization of 1-methylcyclohexane, hydrogenation of ethylene, cracking of some hydrocarbons, hydration of ethyl ether, polymerization of cycloolefins and Claus reaction (76). Pseudoboehmites can provide a wide variety of alumina products meeting specific criteria for different applications. Thus transition phase aluminas made from pseudoboehmites are conceivably potential candidates as membrane reactors for the aforementioned reactions provided the reaction temperature does not appreciably exceed 400°C. In the process of preparing inorganic membranes, particularly for catalytic reactions, the interaction of the membrane and the reaction should be taken into considerations.

The thermal stability of a membrane catalyst must be taken into account when designing a membrane reactor (42). Carbonaceous deposits may occur from the reactions of hydrocarbons and consequently reduce the catalytic activity of the membrane reactor. Some methods must be devised to reactivate the catalytic membrane and/or maintain open pore paths in the membrane. This can be done, for example, by passing oxygen or air at high temperatures through palladium-based membranes or alumina membranes (60).

The distribution of the catalyst in the membrane, in terms of the position across the membrane thickness or along the membrane length, has not been characterized or studied for its effects in any of the references cited previously. And finally all the available literature data is based on simple catalytic reactions. The feasibility of inorganic membrane reactors needs to be extended to more complex cases involving consecutive or parallel reactions.

End Seals and Module Packings

The ends of membrane tubes or monoliths need to be impermeable to all reaction components to prevent the permeate(s) from remixing with the retentate(s) or the retentates from leaking to the permeate(s) through the ends. The long established technologies of ceramic and glass industries are capable of providing the solution to this problem.

A similar but technically more challenging issue is related to module packaging. For example, how can one find a material for connecting a membrane tube to a solid tube or pipe without suffering from such typical problems as thermal stability, cracking due to difference in thermal expansion properties, etc.? How can one find a flexible, high-temperature gasket material for connecting membrane tubes or monolithic elements to header plates or module housing so that thermal expansion does not pose a problem? High-temperature polymers (e.g., polyimides, polyetherimides and polyamideimides which are known to withstand a temperature >230°C for an extended period), gland packing and ceramic to metal seals are just some possibilities.

CONCLUSIONS

Inorganic membranes are reviewed for their potentials as or in membrane reactors. Continuous removal of a product or products from a reaction mixture through an inorganic membrane has been demonstrated to significantly increase the conversion by displacing the reaction equilibrium toward the product side, and, in some cases, enhance the selectivity by suppressing undesirable side reactions. Controlled addition of a second, membrane-selective reactant through an inorganic membrane can also result in improved conversion and selectivity.

The preparation and microstructural as well as separation characteristics of commercial and experimental inorganic membranes are briefly reviewed in the context of their potential uses in the membrane reactor technology. Palladium-based dense membranes have been studied extensively, and practiced industrially to some extent, for hydrogenation- and dehydrogenation-type reactions. Other dense inorganic membranes evaluated in the past include silver and zirconia. Catalyzed porous membrane reactors made from alumina, glass and other oxides have been investigated for various reactions such as decomposition of hydrogen sulfide, dehydrogenation of cyclohexane, dehydration of isopropanol, (oxidative) dehydrogenation of methanol, decomposition of RuO_4 and conversion of sucrose to glucose. In contrast to the solution-diffusion mechanism for dense membranes, the prevailing separation mechanism in those porous inorganic membranes studied in the literature is Knudsen diffusion.

Based on the literature data, basic attributes for effective membrane reactors, such as selective permeability, thermal and structural stabilities, catalytic activity and reaction rate, are discussed together with some engineering aspects of membrane reactors.

Key technological issues essential to further developments of inorganic membranes in or as membrane reactors are identified.

LITERATURE CITED

1. Meindersma, G. W., Procestechniek, 41, 107 (1986).
2. Hsieh, H. P., AIChE Symp. Ser., No. 261, Vol. 84, Am. Inst. Chem. Engr., 1 (1988).
3. Hsieh, H. P., R. R. Bhave and H. L. Fleming, J. Membrane Sci., 39, 221(1988).
4. Shinji, O., M. Misono and Y. Yoneda, Bull. Chem. Soc. Japan, 55, 2760 (1982).
5. Roth, J. F., "Future Catalysis for the Production of Chemicals," in Catalysis 1987, Ward, J. W. (Ed.), Elsevier Sci. Publ., Amsterdam (1988).
6. Gryaznov, V. M., Platinum Metals Rev., 30, 68 (1986).
7. Gillot, J. and A. Auriol, French Patent 2,549,736 to Ceraver (1985).
8. Alary, J., L. Cot, F. Gugliermotte, A. Larbot and C. Guizard, U.S. Patent 4,562,021 to Ceraver (1985).
9. Lillo, E., British Patent Application 2,159,729A to Norton (1985).
10. Schleinitz, H. M., U.S. Patent 4,251,377 to Du Pont (1981).
11. Quinn, J. A., J. A. Anderson, W. S. Ho and W. J. Petzny, Biophysical J., 12, 990 (1972).
12. Riedel, C. and R. Spohr, J. Membrane Sci., 7, 225 (1980).

13. Nippon Sheet Glass, Japanese Unexamined Patent Application 60,186,417 to Nippon Sheet Glass (1985).

14. Matsushita Electric Industry, Japanese Unexamined Patent Application 60,129,119 to Matsushita Electric Industry (1985).

15. Suzuki, H., U.S. Patent 4,699,892 to Suzuki (1987).

16. Bein, T., K. Brown, P. Enzel and C. J. Brinker, Proc. Mat. Res. Soc. Spring Meeting, Reno, 144 (1988).

17. Mitsubishi Heavy Ind., Japanese Unexamined Patent Application 61,192,314 to Mitsubishi Heavy Ind. (1986).

18. Koresh, J. E. and A. Soffer, Separation Sci. Technol., 18, 723 (1983).

19. Gillot, J., European Patent 92,840 to Ceraver (1983).

20. McCaffrey, R. R., R. E. McAtee, A. E. Grey, C. A. Allen, D. G. Cummings and A. D. Appelhans, J. Membrane Sci., 28, 47 (1986).

21. Teijin, K. K., Japanese Unexamined Patent Application 59,177,117 to Teijin (1984).

22. Croopnick, G. A. and D. M. Scruggs, U. S. Patent 4,608,319 to Dresser Industries (1986).

23. Yoldas, B. E., Ceram. Bull., 54, 285 (1975).

24. Leenaars, A. F. M., K. Keizer and A. J. Burggraaf, J. Mat. Sci., 19, 1077 (1984).

25. Larbot, A., J. A. Alary, J. P. Fabre, C. Guizard and L. Cot, Proc. Mat. Res. Soc. Symp., 73, 659 (1986).

26. Beaver, R. P. European Patent Application 186,129 to PPG Industries (1986).

27. Smith, A. W., J. Electrochem. Soc., 120, 1068 (1973).

28. Rai, K. N. and E. Ruckenstein, J. Catalysis, 40, 117 (1975).

29. Mitrovic, M. and L. Knezic, Desalination, 28, 147 (1979).

30. Itaya, K., S. Sugawara, K. Arai and S. Saito, J. Chem. Engr. Japan, 17, 514 (1984).

31. Rigby, W. R., R. C. Furneaux and A. T. Thomas, European Patent Application 242,204 to Alcan International (1987).

32. Lee, K. H. and S. J. Khang, Ceram. Engr. Sci. Proc., 8, 85 (1987).

33. Compagnie des Metaux Precieux, French Published Patent Application 2,302,273 to Compagnie des Metaux Precieux (1976).

34. Nagamoto, H. and H. Inoue, Bull. Chem. Soc. Japan, 59, 3935 (1986).

35. Nazarkina, E. B. and N. A. Kirichenko, Khim. Tekhnol. Topl. Masel, 3, 5 (1979).

36. Itoh, N., Int'l Congress Membr. Membr. Processes, Tokyo, Japan, June 8-12 (1987).

37. Gryaznov, V. M., V. S. Smirnov, L. K. Ivanova and A. P. Mishchenko, Dokl. Akad. Nauk SSSR, 190, 144 (1970).

38. Pfefferie, W. C., U.S. Patent 3,290,406 to Engelhard Ind. (1966).

39. Mishchenko, A. P., M. E. Sarylova, V. M. Gryaznov, V. S. Smirnov, M. R. Roshan, V. P. Polyakova and E. M. Savitskii, Izv. Akad. Nauk SSSR, Ser. Khim., 7, 1620 (1977).

40. Lebedeva, V. I., Met. i Splavy Kak Membran. Katalizatory M., 112 (1981). (Chem. Abstract 98(9):71615 p)

41. Gryaznov, V. M., V. S. Smirnov and M. G. Slin'ko, Proc. 5th Int'l Congress Catal., North-Holland Publishing Co., Amsterdam and London, Vol. 2, 1139 (1973).

42. Gryaznov, V. M., Z. Phys. Chem. Neve Folge, 147, 761 (1986).

43. Gryaznov, V. M., N. M. Ermilova, N. V. Orekhova and N. A. Makhota, Geterog. Katal., 5 (Pt. 1), 225 (1983).

44. Graham, T., Phil. Trans. Roy. Soc. (London), 156, 399 (1866).

45. Gryaznov, V. M., A. P. Mischenko, V. P. Polyakova, E. M. Savitsky, V. S. Smirnov and E. V. Khrapova, Novel Metals and Their Applications (Sverdlovsk), 28, 218 (1971).

46. Gryaznov, V. M., V. S. Smirnov and M. G. Slin'ko, Proc. 6th Int'l. Congress Catal., The Chemical Soc., London, Vol. 2, 894 (1977).

47. DiCosimo, R., J. D. Burrington and R. K. Grasselli, U.S. Patent 4,571,443 to Standard Oil (1986).

48. Gul'yanova, S. G., V. M. Gryaznov and S. Kanizius, Analiz Sovrem. Zadach v Tochn. Naukakh, 172 (1973).

49. Gryaznov, V. M., V. I. Vedernikov and S. G. Gul'yanova, Kinet. Katal., 27, 142 (1986).

50. Vedernikov, V. I., M. P. Skobel'tsina and K. A. Meerov, Deposited Doc., VINITI, 5037-84, 89 (1984).

51. Cales, B. and J. F. Baumard, High Temp.-High Pressures, 14, 681 (1982).

52. Otsuka, K, and A. Morikawa, Kagaku Kogaku, 49, 61 (1985).

53. Nigara, Y. and B. Cales, Bull. Chem. Soc. Japan, 59, 1997 (1986).

54. Guther, W. and W. Vielstich, Electrochimica Acta, 27, 811 (1982).

55. Nogerbekov., B. Yu., N. N. Gudeleva and R. G. Mustafina, Tr. Inst. Org. Katal. Elektrokhim., Akad. Nauk Kaz. SSR, 23, 214 (1984).

56. Nakajima, M., N. Jimbo, K. Nishizawa, H. Nabetani, A. Watanabe and S. Nakao, Int. Congress Membr. Membr. Processes, Tokyo, Japan, June 8-12 (1987).

57. Kameyama, T., M. Dokiya, M. Fujishige, H. Yokokawa and K. Fukuda, Int. J. Hydrogen Energy, 8, 5 (1983).

58. Abe, F., European Patent Application 228,885 to NGK Insulators (1987).

59. Furneaux, R. C., A. P. Davidson and M. D. Ball, European Patent Application 244,970 to Alcan International (1987).

60. Zaspalis, V. T., K. Keizer, J. G. van Ommen, W. van Praag, J. R. H. Ross and A. J. Burggraaf, AIChE National Meeting, Denver, CO, August 17-26 (1988).

61. Kameyama, T., M. Dokiya, M. Fujishige, H. Yokokawa and K. Fukuda, I&EC Fund., 20, 97 (1981).

62. Sun, Y. M. and S. J. Khang, I&EC Res., 27, 1136 (1988).

63. Peng, N., F. Wang and D. Zhou, Fushe Fanghu (Radiation Protection), 3, 154 (1983).

64. Itoh, N., Y. Shindo, T. Hakuta and H. Yoshitome, Int. J. Hydrogen Energy, 9, 835 (1984).

65. Itoh, N., Y. Shindo, K. Haraya, K. Obata, T. Hakuta and H. Yoshitome, Int. Chem. Eng., 25, 138 (1985).

66. Asaeda, M. and L. D. Du, J. Chem. Eng. Japan, 19, 72 (1986).

67. Lillo, E., U.S. Patent 4,603,109 (1986).

68. Baker, R. A., G. D. Forsythe, K. K. Likhyani, R. E. Roberts and D. C. Robertson, U.S. Patent 4,105,548 to Du Pont (1978).

69. Gryaznov, V. M., USSR Author's Certificate 274,092 (also British Patent 1,234,855) to V. M. Gryaznov (1964).

70. Uhlhorn, R. J. R., K. Keizer and A. J. Burggraaf, to appear in ACS Symposium Series (1988).

71. Schnabel, R. and W. Vaulont, Desalination, 24, 249 (1978).

72. Messing, R. A., British Patent Application 2,014,868A to Corning Glass (1979).

73. Shimizu, Y., T. Yazawa, H. Yanagisawa and K. Eguchi, J. Ceram. Soc. Japan, Int'l Ed., 95, 1012 (1987).

74. Shimizu, Y., T. Yazawa, H. Yanagisawa and K. Eguchi, J. Ceram. Soc. Japan, Int'l Ed., 95, 1113 (1987).

75. Yazawa, T., H. Nakamichi, H. Tanaka and K. Eguchi, Nippon Seramikkusu Kyokai Gakujutsu Ronbunshi, 96, 18 (1988).

76. Oberlander, R. K., "Aluminas for Catalysts - Their Preparation and Properties," in Applied Industrial Catalysis (Vol. 3), Academic Press, New York (1984).

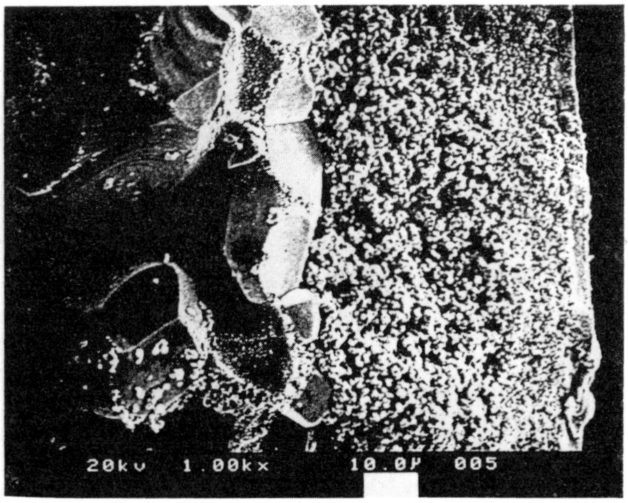

Figure 1. Scanning electron micrograph of a multi-layered alumina membrane-support composite.

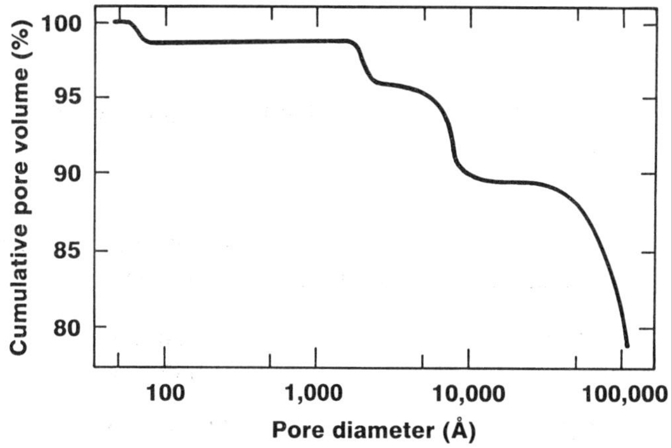

Figure 2. Pore size distribution of a multi-layered alumina membrane-support composite.

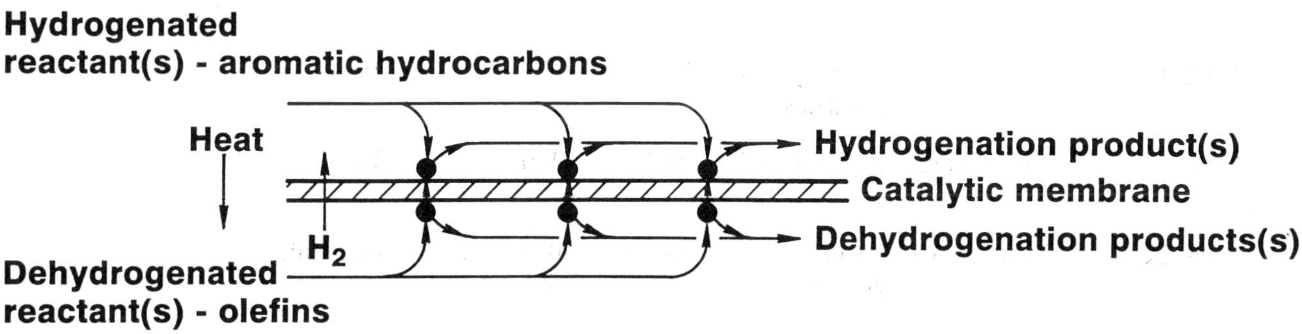

Figure 3. Coupling of two reactors partitioned by a membrane permeable only to a common reaction component.

Table 1. General Advantages of Membrane Reactors Over Fixed/Fluidized Bed Reactors

1. Controlled pore size and distribution
2. Controlled pore direction
3. Controlled contact time
4. More effective fluid/catalyst contact
5. High surface/volume ratio
6. Less mass transfer resistance
7. Reaction/separation in one unit operation
8. Less stagnant zones - less catalyst fouling
9. Easier to reactivate catalysts
10. High rate of heat transfer

Table 2. Commercial Porous Inorganic Membranes

Manufacturer	Trade Name	Membrane Material	Support Material	Membrane Pore Diameter	Geometry of Membrane Element	Note
Mott		SS, Ni, Au, Ag, Pt, etc.		≥0.5 μm	Tube	
Pall		SS, Ni, etc.		≥0.5 μm	Tube	
Osmonics	Hytrex®	Ag		0.2-5 μm	Disk	
Union Carbide	Ucarsep®	ZrO_2	C	40 Å	Tube	Distributor: Gaston County Dyeing Machine Co.
SFEC	Carbosep®	ZrO_2 ZrO2	C C	~40 Å 0.08-0.14 μm	Tube	Modified Ucarsep® technology
CARRE		ZrO_2 (dynamic membrane)	SS		Tube	
TDK	Dynaceram®	ZrO_2 (dynamic membrane)	Al_2O_3	~100 Å	Tube	
Alcoa/SCT	Membralox®	Al_2O_3 Al_2O_3	Al_2O_3 Al_2O_3	40-1000 Å 0.2-5 μm	Monolith/ Tube	
Norton	Ceraflo®	Al_2O_3	Al_2O_3	0.2-1.0 μm	Monolith/ Tube	
NGK		Al_2O_3	Al_2O_3	0.2-5 μm	Tube	
Anotec/Alcan	Anopore®	Al_2O_3 Al_2O_3	Al_2O_3 Al_2O_3	250 Å 0.2 μm	Disk	Made by anodic oxidation
Asahi Glass		Glass Glass		40-500 Å ≤5 μm	Tube/Disk	
Schott Glass		Glass		100 Å 0.1 μm	Tube	
Fuji Filters		Glass Glass		40-900 Å 0.25-1.2 μm	Tube	
Ceram Filters		SiC	SiC	0.15-8 μm	Tube	

Table 3. Experimental Inorganic Membranes

Membrane Material	Reference	Note
TiO_2	7	
	8	Sol-gel process
	9	
Mullite	9	
Cordierite	9	
	10	
Mica	11	Track-etch process; straight pores
	12	
Cr_2O_3	9	
MgO	9	
SnO_2	13	
Zeolite	14	
	15	
	16	
	17	
Molecular sieve carbon	18	Pyrolysis process
SiC	9	
	19	
Si_3N_4	19	
TiC	9	
Polyphosphazene	20	Inorganic polymer
Transition metals or alloys	21	Deposition process
Amorphous metals or alloys	22	

Table 4. Examples of Literature Studies of Gas Separation by Inorganic Membranes

1. Dehumidification of air (alumina membrane)
2. He/CO_2 separation (glass membrane)
3. O_2/N_2, He/N_2, He/O_2, N_2/SF_6 (molecular sieve carbon membranes)
4. O_2/H_2O (metal oxide membranes)
5. H_2/H_2S (glass and alumina membranes)
6. N_2/H_2, N_2/C_3H_6 (molecular sieve carbon membranes)
7. Alcohol/H_2O (alumina and alumina/silica membranes)
8. H_2/N_2, H_2/CH_4 (alumina/titania/silica membrane)
9. O_2/N_2 (zeolite membranes)
10. H_2/N_2 (alumina/titania membrane)
11. H_2/N_2, H_2/CH_4 (alumina membrane)
12. Isotopic enrichment of uranium (alumina/zirconia membrane)*

*Used in industrial-scale process.

Table 5. Examples of Liquid Separation Applications of Inorganic Membranes

1. Oil/water emulsion separation
2. Textile sizing recovery
3. Sterilization and clarification of fruit juices
4. Extraction of serum proteins from whey
5. Production of cream cheese from curdled milk
6. Clarification of wine and beer
7. Purification of radioactive laundry water
8. Metal cleaning bath processing
9. Recovery of lignin from bleached kraft pulp process

Table 6. Catalytic Dense Palladium and Its Alloy Membranes

Membrane Material	Catalytic Reaction	Reference
Pd	Decomposition of water to hydrogen	33
	Hydrogenation of butadiene	34
	Steam reforming of methane	35
	Dehydrogenation of cyclohexane	36
	Dehydrogenation of butene to butadiene	37
Pd-Ag	Dehydrogenation of ethane	38
Pd-Cu	Dehydrogenation of 1,2-cyclohexanediol	39
Pd-Mo	Hydrodemethylation of methyl- and dimethylnaphthalenes	40
Pd-Ni	Hydrodealkylation of dimethylnaphthalene	41
Pd-Rh	Dehydrogenation of n-heptane to benzene and toluene	6
Pd-Ru	Hydrogenation of pentadiene	42
Pd-Sb	Hydrogenation of butenes	43

Table 7. Porous Inorganic Membranes Studied as Membrane Reactors

Membrane Material	Catalyst	Reaction	Temperature	Reference	Note
Al_2O_3	Enzyme immobilized	Conversion of sucrose to glucose		56	Conversion increased from 50% to ~100%
Al_2O_3	MoS_2 in packed bed	Decomposition of H_2S to form H_2 and S	800°C	57	Conversion twice as the reaction equilibrium conversion
Al_2O_3	MoS_2 in membrane matrix	Decomposition of H_2S to form H_2 and S	800°C	58	Conversion twice as the reaction equilibrium conversion
Al_2O_3	Al_2O_3	Dehydration of isopropanol	200°C	59	Conversion to propene and water
Al_2O_3	Al_2O_3	Oxidative dehydrogenation of methanol	20-500°C	60	Reaction rate ten times as in fixed bed reactors
Glass	MoS_2 or WS_2 in packed bed	Decomposition of H_2S to form H_2 and S	800°C	61	Conversion twice as the reaction equilibrium conversion
Glass	Pt/Al_2O_3 in packed bed	Dehydrogenation of cyclohexane	170-360°C	4	Conversion increased from 33% to 80%
Glass	Pt/Al_2O_3 in packed bed	Dehydrogenation of cyclohexane	180-220°C	36	Conversion increased from 18% to 45%
Glass	Platinum impregnated	Dehydrogenation of cyclohexane	300°C	62	Conversion increased from 61% to 90%
Fe_2O_3/ molecular sieves/clay	Fe_2O_3 in membrane matrix	Decomposition of RuO_4 to RuO_2 and O_2	500°C	63	Conversion reached 99.4%

STABILITY OF H_2-PERMSELECTIVE SiO_2 FILMS FORMED BY CHEMICAL VAPOR DEPOSITION

S.W. Nam and G.R. Gavalas ■ Department of Chemical Engineering, California Institute of Technology, Pasadena, CA 91125

Thin SiO_2 films were heat treated in different gas mixtures to determine their stability in functioning as high-temperature hydrogen permselective membranes. The films were formed within the walls of porous Vycor tubes by SiH_4 oxidation in an opposing reactants geometry. Film deposition was carried out at 450 °C in the presence and absence of water vapor. Immediately after formation, the films were highly selective to hydrogen permeation having a H_2:N_2 permeability ratio of about 3000 : 1. The films were subsequently heat treated at 450–700°C in dry N_2, dry O_2, N_2–H_2O, and O_2–H_2O mixtures. The permeation rates of H_2 and N_2 changed depending on the original conditions of film formation as well as on the heat treatment. Heating in dry N_2 reduced slowly the permeation rates of both H_2 and N_2. Heating in N_2–H_2O atmosphere led to a steeper decline of H_2 permeability. But the permeation rate of N_2 increased or decreased according to whether the film deposition had been carried out in the absence or presence of H_2O vapor, respectively. Thermal treatment in O_2 caused rapid decline of the permeation rates of H_2 and N_2 in films that were deposited under dry conditions. The decline was moderate in films deposited under wet conditions.

INTRODUCTION

Hydrogen permselective ceramic membranes can be used in membrane-reactor configurations to improve the yield of certain equilibrium-limited catalytic reactions such as hydrogen sulfide decomposition and water gas shift reaction. Such combination of catalysis and gas separation requires a membrane with high selectivity, adequate permeation rate and mechanical, thermal and chemical stability. Amorphous silicon dioxide in nonporous form has very high selectivity but low permeability for hydrogen. Its application to membrane separations consequently hinges on the ability to produce it in the form of a very thin film free of cracks or pinholes and applied on a porous support material to achieve mechanical stability.

We have recently deposited films of amorphous SiO_2 within the walls of porous Vycor tubes by SiH_4 oxidation in an opposing reactant geometry (1): SiH_4 was passed inside the tube while O_2 was passed outside the tube. The two reactants diffused opposite to each other and reacted within a narrow front inside the tube wall to form a thin SiO_2 film. Measurements of H_2 and N_2 permeation rates showed that the SiO_2 film was highly selective to H_2 permeation. The H_2:N_2 flux varied between 2000-3000 while the hydrogen permeation coefficient was on the order of 0.2 cm^3(STP)/cm^2-atm-min.

Silicon dioxide formed by chemical vapor deposition is known to undergo densification upon prolonged exposure to high temperatures (2,3). This densification is intensified in the presence of water vapor to such an extent that heating in steam at 450°C for 10 hours decreased the film thickness by 7.9% from the as-deposited value (4). In our previous experiments we found that thermal treatment in the presence of water vapor caused significant changes in the film properties (1). In particular we observed a decrease in the hydrogen permeation coefficient and selectivity of the membrane.

In membrane reactor operation, the extent of property degradation would depend on the gaseous environment (reactants, products) and the temperature. For example, in hydrogen sulfide decomposition the membrane will be subjected to temperatures 600°C or higher, while in the water gas shift reaction the membrane will be exposed to water vapor but at substantially lower temperatures. The purpose of this paper was to report some results on the effect of thermal treatment at different temperatures and gaseous atmospheres. Some limited results are also reported on the effect of water vapor during film deposition on the subsequent film stability to thermal treatment.

EXPERIMENTAL

Film Deposition

Films of SiO_2 were deposited within the walls of porous Vycor tubes by SiH_4 oxidation in the opposing reactant geometry described elsewhere (1). The dry deposition was carried out at 450 and 440°C with 0.05 atm of SiH_4 in N_2 and 0.15 atm of O_2 in N_2. The humid deposition was carried out at the same temperatures with 0.05 atm of SiH_4 in N_2, and 0.04 atm of H_2O and 0.15 atm of O_2 in N_2. The moist gases were prepared by passing the oxygen/nitrogen stream through a glass vessel filled with water and maintained at 30°C. The deposition reactions were carried out for 30 to 60 minutes until the rates of N_2 permeation decreased to about 10^{-5} cm^3(STP)/cm^2-atm-min.

Heat Treatment

The tubes bearing the deposited SiO_2 films were subjected to various heat treatments. The ambient gases used were dry N_2, 4% H_2O in N_2 (wet N_2), dry O_2, and 4% H_2O in O_2 (wet O_2). One treatment involved heating of the film deposited under dry condition at 450°C with dry N_2, wet N_2, and dry O_2. In another treatment, the tubes were successively heated in different gaseous environments for 1 day at 440, 600, and 700°C.

Permeability Measurements

The rates of hydrogen and nitrogen permeation through the tubes during the reaction and during various treatments were measured in the concentric system used for the deposition experiments. Hydrogen or nitrogen was passed through the inside of the tube usually at 310 kPa pressure while the system was maintained at 440 and 450°C. The rate of gas permeation to the annular space was measured by a bubble flow meter when the gas flux was high. To measure lower permeation rates, the annular space was evacuated and connected to a Baratron pressure gauge interfaced with a microcomputer. After the annular volume was isolated from the evacuation system the initial pressure increase gave directly the permeation rate with the use of a calibration curve. The permeation coefficient (cm^3 (STP)/cm^2-atm-min) was defined as the flux of gas through the inner surface area of the tube when the pressure difference was 1 atm.

RESULTS AND DISCUSSION

Effects of water during film deposition

The evolution of hydrogen and nitrogen permeation rates after successive 5-minute reaction intervals is shown in Figure 1. Before deposition, the H_2-N_2 flux ratio was close to 3.7, the ratio of Knudsen diffusivities. During SiO_2 deposition the permeation rates of both gases decreased. In 30 minutes the H_2 permeation rate declined from 0.7 to 0.2 cm^3 (STP)/cm^2-atm-min whereas the N_2 permeation rate decreased from 0.2 to 10^{-5} cm^3(STP)/cm^2-atm-min. With water vapor present during the reaction, the permeation rate of N_2 declined further. Comparing two films when they have the same value of H_2 permeation rate, the N_2 permeation rate through the film deposited in the presence of water is lower than the rate through the film deposited under dry conditions (Fig. 2). The somewhat higher H_2 selectivity of the film prepared in the presence of water vapor may be due to the formation of a denser but more uniform film.

Thermal treatment

The first series of densification experiments was carried out at 450°C under dry N_2 and wet N_2 (4% H_2O in N_2) on a film that had been deposited under dry conditions. The permeation rates of H_2, N_2, and He during a 5-day treatment period are presented in Figure 3. The permeability plots demonstrate that treatment with wet N_2 causes a pronounced change in the permeation rates. Thermal treatment in wet N_2 for 1 day decreases the H_2 flux by 55% and the decrease becomes 68% after five days. The permeation rate decreases rapidly during the first day of treatment and then decreases more slowly. When the heat treatment is carried out in dry N_2, the H_2 flux decreased by only 1% or 9% after one day and five days respectively. The flux of N_2 also changed with thermal treatment. In dry N_2, the flux of N_2 decreased by 32% in 5 days while in treatment with wet N_2, the N_2 flux decreased by 83% in the same period. The permeation rate of He did not change very much upon thermal treatment. Treatment for 5 days in dry N_2 and wet N_2 caused 5% and 25% declines in the He permeation rate.

The second series of thermal treatment experiments was carried out under dry N_2, wet N_2, dry O_2, or wet O_2 using films deposited under dry or wet conditions. The temperature was stepwise increased from 440°C to 600°C to 700°C with 1 day heating at each temperature. The gas flux generally decreased with each treatment and the rate of this decrease depended on both ambient conditions and film type.

For a film that had been deposited under dry conditions, treatment with dry N_2 at 440°C caused a small change in permeation rates as mentioned earlier. But treatment with dry N_2 at 600°C significantly decreased the H_2 flux and further heating at 700°C reduced this flux further as shown in Figure 4. The flux of N_2 also decreased slowly under heating in dry N_2. The decline of the H_2 and N_2 permeabilities reflects the densification of the SiO_2 films. In the case of H_2, densification raises the activation energy of permeation as shown in the Arrhenius plot of Figure 5. The activation energies estimated for films treated in dry N_2 at 440, 600, and 700°C were 6.2, 22, and 30 kJ/mole, respectively.

Heat treatment in wet N_2 at 440°C of films deposited under dry conditions decreased the H_2 flux and increased the N_2 flux, as mentioned earlier. Further heating at 600 and 700°C with wet N_2 reduced the H_2 flux and increased the N_2 flux further. But once the film was subjected to densification at 440°C, the change of the gas flux upon further heating was relatively small. Heat treatment in dry O_2 significantly decreased the permeation rates of both H_2 and N_2, although the decrease was more pronounced for N_2. These decreases are rapid in the first 6 hours at 440°C and then become very slow. Further heating under O_2 at 600°C caused a slow further decline in the permeation rate of H_2 and N_2. However, heat treatment under O_2 at 700°C caused further reduction of the H_2 rate but increased the N_2 rate to the level measured immediately after deposition.

The results of thermal treatment of films deposited under wet conditions are shown in Figure 6. The N_2 and H_2 permeation rates decreased with each treatment. The H_2 permeation rate decreased to a lesser extent than for the case of the film deposited under dry conditions. The permeation rate of N_2 declined with wet N_2 treatment unlike the experiments with the film deposited under dry conditions. Wet O_2 treatment caused reduction of the N_2 permeation rate even at 700°C, again unlike the experiments with films produced under dry conditions.

As mentioned earlier, the changes in gas permeabilities are caused by the densification of the SiO_2 film material. This densification takes place by thermal treatment alone but is more pronounced in the presence of water vapor (2,3). It has been suggested that densification by thermal treatment involves removal of silanol groups (Si-OH) in the SiO_2 film and formation of Si-O-Si bonds to close the channel. Water vapor seems to strongly catalyze the rearrangement of siloxane bonds and help dehydroxylation and thus densification (5).

The increase of N_2 permeation with wet N_2 treatment of films formed under dry conditions suggests microcrack formation during densification. Deposition under dry conditions may create tensile stresses in a SiO_2 film but subsequent thermal treatment of the stressed film with dry N_2 may cause uniform densification avoiding crack formation. In the absence of cracks, densification reduces the permeation rates of both H_2 and N_2. Wet N_2 treatment, on the other hand, decreases the permeation of hydrogen but increases the permeation of nitrogen. The reasons for this opposite direction of the change in N_2 permeation rate caused by dry and wet thermal treatments are unclear. One possible explanation can be sought in the uneven infiltration of water in the film creating regions of uneven shrinkage which results in the formation of microcracks. The formation of cracks in SiO_2 films was cited previously (6). Deposition on silicon and even quartz results in cracking especially when the film is thick. H_2 permeation through these microcracks makes a relatively minor contribution in view of the dominant contribution of diffusion through the nonporous material. But diffusion of nitrogen through the microcracks can become the major contribution in view of the negligible permeation through the nonporous material.

It has been found that the stress in CVD SiO_2 films could be decreased by the addition of water vapor to the reactants(7). For films formed in the presence of water vapor, the decrease rather than increase of the N_2 permeation rate suggests that microcracks are not generated during the thermal treatment of this material. One possible explanation of this behavior is that water vapor during the reaction relieves the stresses in the film allowing uniform densification upon subsequent thermal treatment.

The rapid decrease in permeability caused by heating in O_2 at 440°C can again be attributed to film densification. This finding is consistent with the results of a previous study showing that densification at 770°C was more rapid under air than under Ar (4). The SiO_2 films tested in this latter study were prepared by SiH_4 oxidation at O_2:SiH_4 ratios above 10. When this ratio fell below 4, the films were deficient in oxygen and contained an intermediate stoichiometric oxide Si_2O_3 (8) or a nonstoichiometric phase (9). The O_2:SiH_4 ratio in our experiments varied spatially within the porous wall from extremely oxygen-rich on the one side of the tube to extremely silane-rich at the other side. It is likely therefore that the films contained oxygen deficient regions which were oxidized and became denser during subsequent thermal treatment.

Evolution of film permeability during alternating periods of deposition and thermal treatment

A final experiment was carried out consisting of alternating intervals of deposition at 440°C in the presence of water vapor and six-hour intervals of thermal treatment at 600°C under wet O_2. After each thermal treatment the temperature was decreased to 440°C and the permeation rates of H_2, N_2, and He were measured at that temperature. Figure 7 shows the evolution of gas flux as a function of cumulative deposition time. In 5 minutes of deposition, the flux of N_2 decreased drastically, while the fluxes of H_2 and He underwent a modest decline. Interestingly, the He:H_2 ratio which was below unity before deposition increased to 2 after 5 minutes of deposition. After 15 minutes deposition (cumulative) and intervening periods of heat treatment, the H_2:N_2 ratio increased to 2800. After the last interval of deposition (15 minutes cumulative) and the last heat treatment period, the permeabilities of the three gases were measured at different temperatures. An Arrhenius plot of these measurements is shown in Figure 8. The N_2 permeation rate is very low and follows the Knudsen temperature dependence ($\alpha\ T^{-\frac{1}{2}}$). The permeation rates of H_2 and He are activated with energies 26.4 and 6.8 kJ/mole respectively.

CONCLUSIONS

Films of SiO_2 deposited within the walls of porous Vycor tubes by dry or wet SiH_4 oxidation have been subjected to various thermal treatments which caused changes in permeation rates of H_2 and N_2. These changes depended on treatment temperature, gaseous atmosphere and film type. Thermal treatment with dry N_2 reduced slowly the permeation rates of both H_2 and N_2. Treatment with H_2O-N_2 mixtures significantly decreased the flux of hydrogen. The same treatment increased the flux of nitrogen for those films which were deposited under dry conditions and decreased the flux of nitrogen for the films deposited under wet conditions. Thermal treatment in oxygen had an even larger effect on the permeation rates. Heating under oxygen at temperatures as low as 440°C caused a rapid decrease in the permeation of both hydrogen and nitrogen. Deposition in the presence of water vapor followed by thermal treatment under H_2O-O_2 gave dense and relatively stable films with H_2:N_2 flux ratios above 2000.

ACKNOWLEDGMENT

This work was supported by the Caltech Consortium in Chemistry and Chemical Engineering.

LITERATURE CITED

1. Gavalas, G. R., C. Megiris and S. W. Nam, Chem. Eng. Sci., in press (1989).
2. Swaroop, B., "Low temperature densification of silica films," in Thin Film Dielectrics, Vratny (Ed.), The Electrochem. Soc., New York (1969).
3. Pliskin, W. A., J. Vac. Sci. Technol., 14, 1064 (1977).
4. Kern, W., RCA Rev., 37, 55 (1976).
5. Iler, R. K., "The Chemistry of Silica," John Wiley and Sons, New York (1979).
6. Goldsmith, N., and W. Kern, RCA Rev., 28, 153 (1977).
7. Kern, W., G. L. Schnable, and R. B. Fisher, RCA Rev., 37, 3 (1976).
8. Ritter, E., Opt. Acta., 19, 197 (1962).
9. Nakamura, M., Y. Mochizuki, K. Usami, Y. Itoh, and T. Nozaki, J. Electrochem. Soc., 132, 482 (1985).

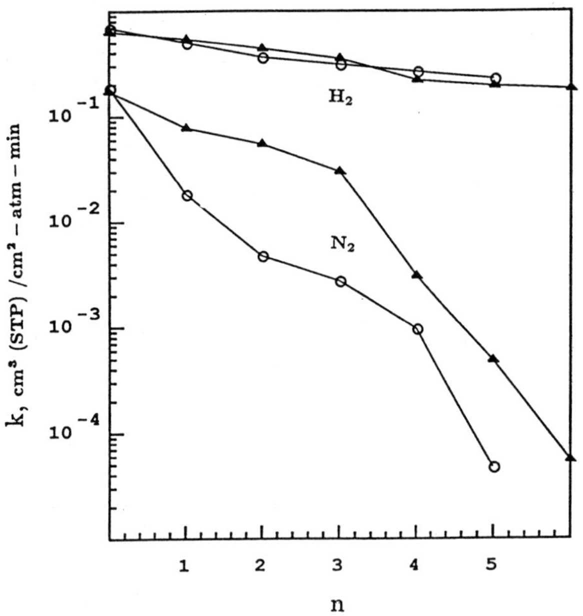

Figure 1. Evolution of H_2 and N_2 permeation rate coefficients at 450°C after successive 5-minute deposition intervals. k : permeation rate coefficient ; n : number of deposition intervals : Δ : dry reaction ; O : humid reaction.

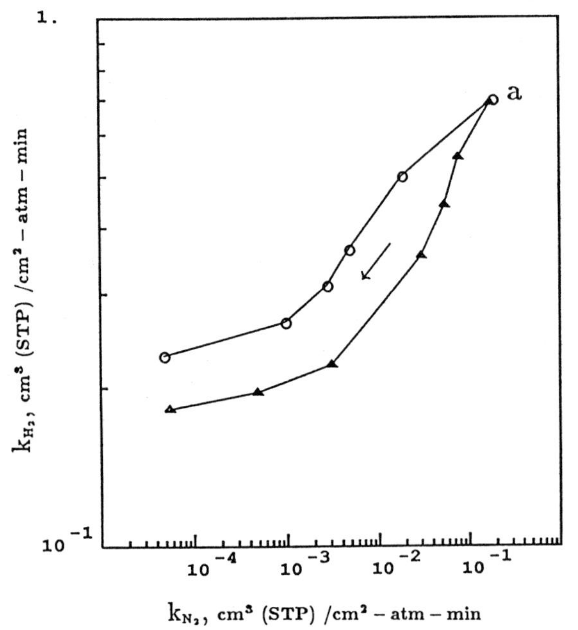

Figure 2. Evolution of H_2 and N_2 permeation rate coefficients at 450°C after successive 5-minute deposition intervals. Arrow indicates the direction of deposition reaction. k_{H2} : permeation rate coefficient of H_2 ; k_{N2} : permeation rate coefficient of N_2 ; a : permeation rate coefficients before deposition ; Δ : dry reaction ; O : wet reaction.

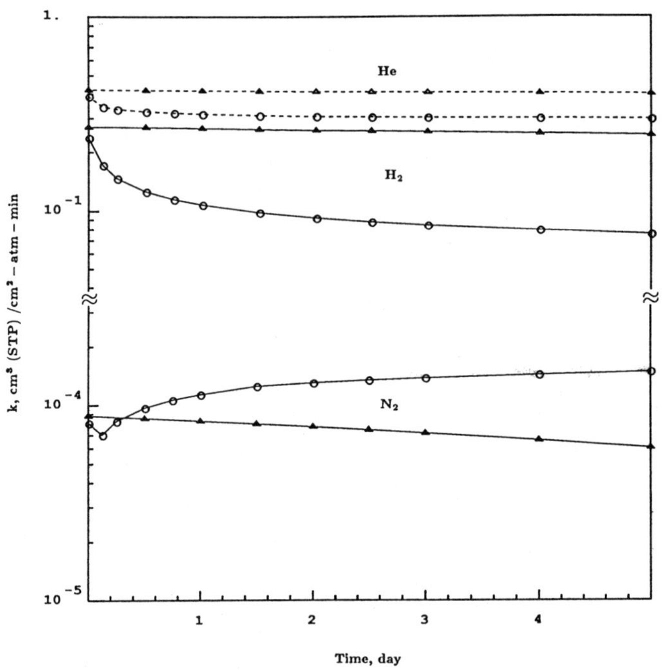

Figure 3. Permeation rate coefficients of gases during a 5-day thermal treatment periods at 450°C. k : permeation rate efficient ; Δ : treatment without H_20 ; O : treatment with H_20.

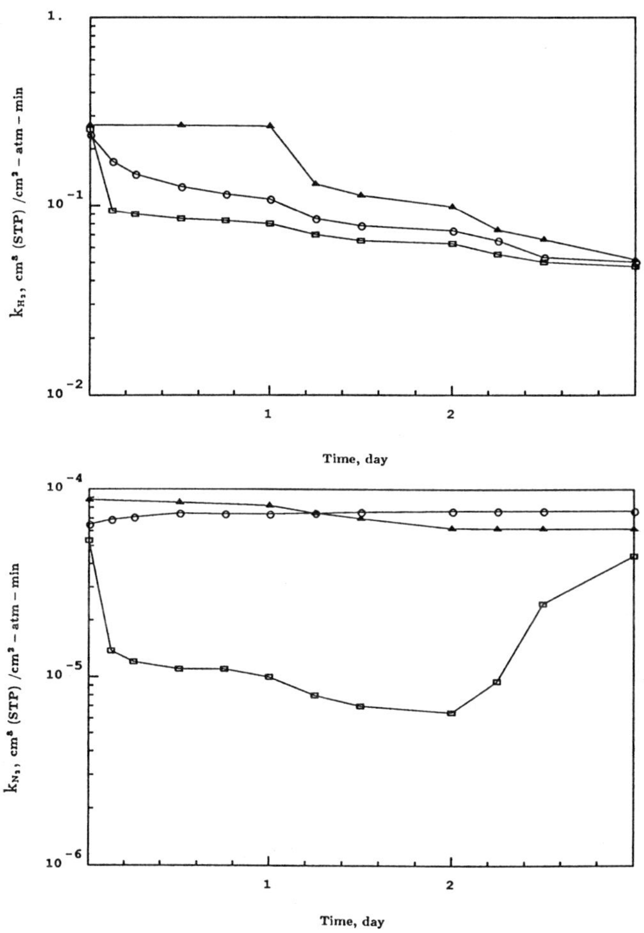

Figure 4. Permeation rate coefficients at 440°C of gases during successive 1-day thermal treatments at 440, 600, and 700°C. Films were deposited under dry conditions. Δ : treatment with dry N_2 ; O : treatment with wet N_2 ; □ : treatment with dry O_2.

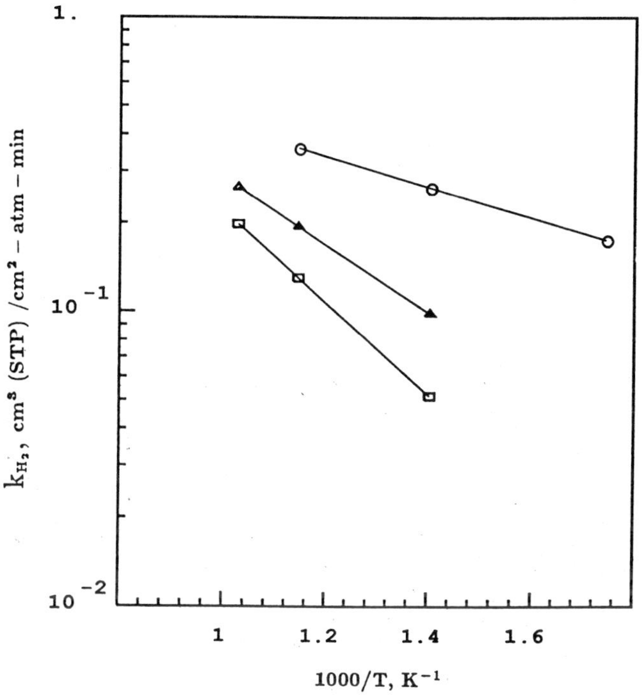

Figure 5. Permeation rate coefficients of H_2 during thermal treatment. Films were deposited under dry conditions. O : after 1 day treatment at 440°C ; Δ : after 1 day treatment at 600°C ; □ : after 1 day treatment at 700°C.

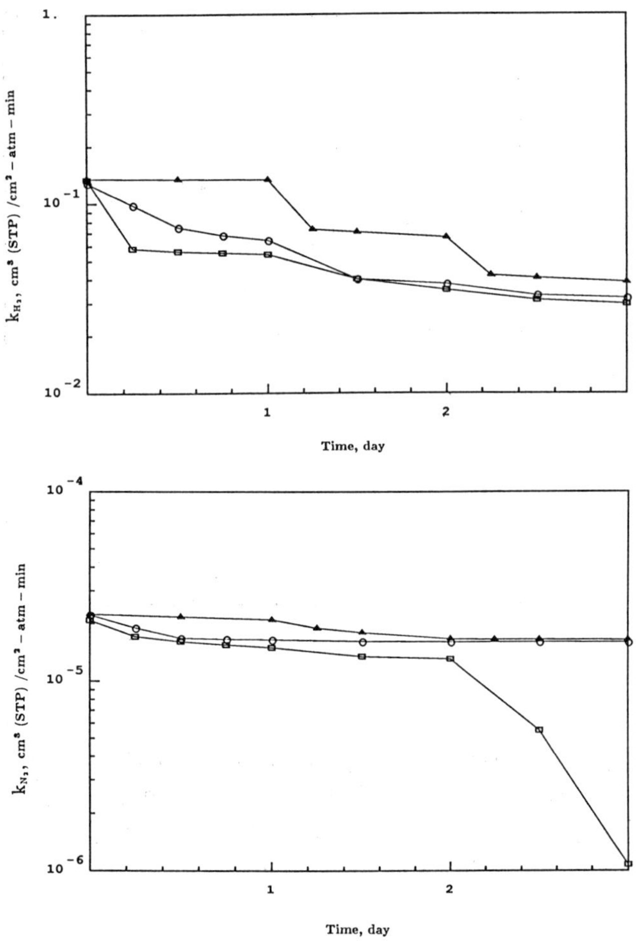

Figure 6. Permeation rate coefficients at 440°C of gases during successive 1-day thermal treatments at 440, 600, and 700°C. Films were deposited under wet conditions. Δ : treatment with dry N_2 ; O : treatment with wet N_2 ; □ : treatment with wet O_2.

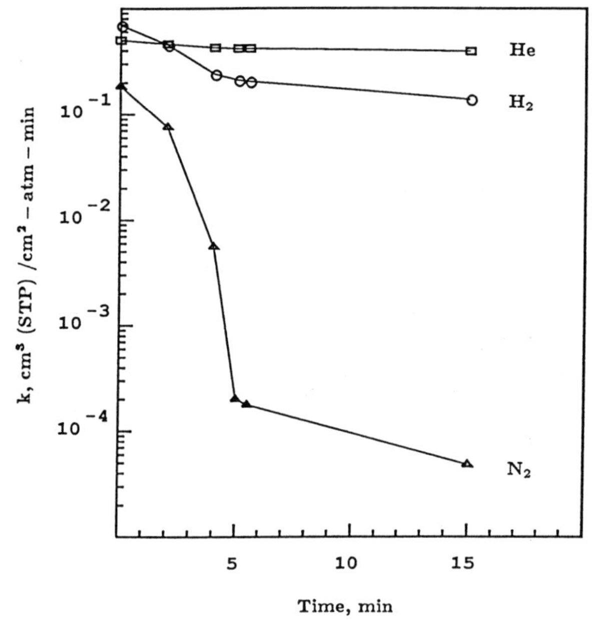

Figure 7. Evolution of permeation rate coefficients at 440°C during alternating periods of deposition and thermal treatment.

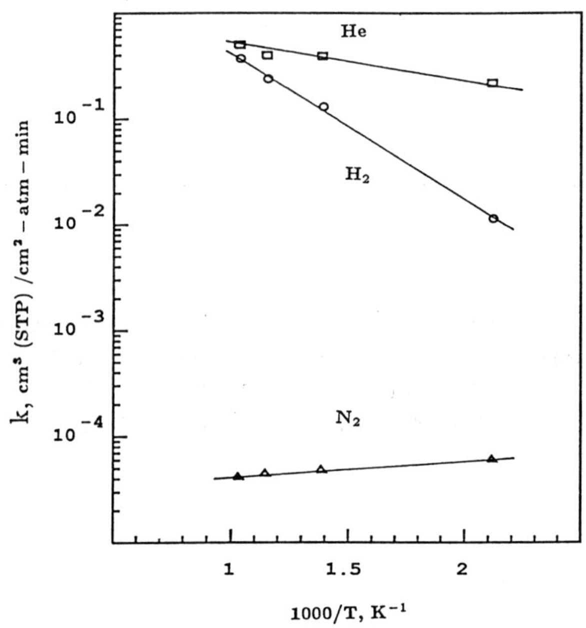

Figure 8. Permeation rate coefficients of H_2, N_2, and He vs. temperature.

TRANSPORT AND REACTION STUDIES USING CERAMIC MEMBRANES

N. Nourbakhsh, A. Champagnie, and T.T. Tsotsis ■ Department of Chemical Engineering, University of Southern California, Los Angeles, CA 90089-1211

I.A. Webster ■ UNOCAL Corporation, Corporate Division, 1201 West 5th Street, P.O. Box 7600, Los Angeles, CA 90051

Ceramic membranes have been available since the early seventies but due to problems with production techniques, which resulted in membranes of poor quality, had not found any significant industrial applications. Improvements in production techniques (1-4) have recently resulted in commercial membranes of high quality in terms of pore size uniformity and with good thermal and mechanical properties and have made such membranes desirable and competitive for a number of industrial applications in the food, pharmaceutical and electronics industries currently utilizing polymeric, organic and metal membranes.

Ceramic membranes also show potential for high temperature catalytic reaction engineering applications particularly for reactions with low conversions and yields due to either unfavorable thermodynamics, product inhibition and undesirable side reactions. Typical reaction systems amenable to the application of ceramic membranes include (the list is by no means exhaustive) hydrogenation and dehydrogenation reactions, hydroformylations, hydrations and dehydrations, olefin metathesis and partial oxidations. Catalytic membrane reactors have the inherent capability for combining reaction and separation in a single unit operation, by itself a significant improvement over the conventional two units operations with the reaction and separation units completely separated. They offer, however, a number of other distinct advantages over traditional reactors (packed, fluidized and trickle beds). In a number of cases the membrane provides for selective removal of one or more products and/or stable intermediates in parallel with the reaction thus driving the reaction - to use equilibrium limited reactions as an example - continuously towards the product side and resulting (in principle as well as in practice) in higher reactor conversions. In other instances the membrane allows one to adjust and influence the surface concentration of the various reactant and intermediate species and therefore provides the flexibility for unique control of the reactor selectivity and product distribution.

The earlier efforts in catalysis and reaction engineering to use membrane reactor technology involved the use of glass (5-8) and Pd membranes (9-14). The development of Pd catalytic membranes, in particular, has had a significant impact on proving and popularizing the concept of catalytic membrane reactors. The low mechanical strength of the glass membranes and the low permeabilities, and high cost of the Pd membranes combined with phenomena of metal sintering and fatigue have so far hindered the widespread industrial application of these materials. Other early membrane reactor efforts involved the use of ceramic materials exhibiting enhanced oxygen anionic conductivity (15-17).

Alumina ceramic membranes, with which we will primarily deal with in this paper, have been synthesized by a variety of techniques such as phase separation and leaching (18), slipcasting (19), Sol-Gel (20-22), anodic oxidation (23-26), track etching (27,28), direct pyrolysis (29,30) and thin film deposition (31,32). They have so far been tested successfully in various gas-phase separations (33-37), for water purification and sterilization operations in the food and pharmaceutical industries (38-40) for the treatment of oily wastewaters (41) and in a number of catalytic membrane reactor applications (42-45).

RESEARCH USING CERAMIC MEMBRANES

Our research group has been working on various applications of high temperature ceramic membranes for the last three years (see Table 1 for an outline of past and ongoing research activities). Our research efforts have had two distinct goals: (a) First to test various ceramic membrane preparation methods and the applicability of such membranes in catalytic and reaction engineering applications, and (b) to gain a better understanding of the various transport and reaction phenomena occuring in such membranes.

Table 1. Outline of past and ongoing studies at USC of transport and reaction phenomena using unimodel membranes.

ITEM	RESEARCH AREA 1	RESEARCH AREA 2
RESEARCH FIELD	PREPARATION CHARACTERIZATION AND AGING OF CERAMIC ALUMINA MEMBRANES	EVALUATION OF TRANSPORT AND REACTIVE PROPERTIES OF PORPHYRINS AND ASPHALTINES IN MEMBRANES
KEY SUCCESSES	• CONTROL OF PORE DIAMETER AND LENGTH • STUDY OF METAL IMPREGNATION • USE OF ALUMINA CERAMIC MEMBRANES IN GAS AND VAPOR PHASE MEMBRANE REACTORS • STUDY OF MECHANICAL AND THERMAL PROPERTIES AND AGING IN BENCH SCALE HYDROTREATERS • KEY PUBLICATIONS: SEE REFS 23,24,42,43,46	• ASPHALTENES DELAMINATE AND DIFFUSE AS MOLECULAR WEIGHT CLASSES. TRANSPORT MODEL DEVELOPED • SURFACE DIFFUSION IS IMPORTANT • KEY PUBLICATIONS. SEE REFS 47-50

Anodic Alumina Membranes

Of the alumina membranes prepared by various methods (see above) only anodic membranes are ideally suited for fundamental studies of transport and reaction, since they have ideal porous structures with vertical nonintersecting pores, surrounded by hexagonal cells (see Fig. 1 for an idealized structure of an anodic porous alumina film). In contrast ceramic membranes prepared by other techniques have highly tortuous porous structures (Sol-Gel membranes for example consist of stacked particles, see (20-22) for further discussion). Alumina membranes have been prepared in our laboratories by the electrochemical anodization of aluminum in various acidic solutions. The pore size, pore density and thickness of these membranes can be precisely controlled by the choice of the electrolyte and by adjusting parameters such as temperature, anodizing voltage and anodization time (further details about our preparation and characterization techniques can be found in a number of our recent publications (23,24,42)). Fig. 2, for example, shows the effect of anodizing voltage on the pore density for films we have prepared in a 0.4M H_3PO_4 solution at 20°C for anodization times of 60 min (24). The trends in the data (i.e., the pore density decreasing with voltage) are consistent with the results of O'Sullivan and Wood (51). XPS scans of these films (24) show corrected binding energies for Al_{2p} (74.6 ev) and O_{1s} (531.9 ev), which are a good match for Al_2O_3.

Studies of Catalyst Deactivation Using Ceramic Membranes

In the last three years we have made extensive use of anodic alumina films as probes for monitoring catalyst deactivation during heavy oil hydroprocessing. Typical hydroprocessing catalysts are metal sulfides supported on high surface area alumina supports. During hydrotreating of coal and crude-derived heavy liquids these catalysts deactivate through carbon and metal deposition. Coke is believed to cause loss in catalytic activity by physical blocking of the pores and by chemical modification of the surface, while metal deposition causes further loss of catalytic activity by active site poisoning and eventually pore mouth plugging. To determine the rate and mechanism of deactivation it is instructive to characterize the spent catalysts for coke and metal contents but the tortuous pore structure of commercial supports and the presence of other metal ingredients greatly complicate this task. Visual observations of how the pore structure of a hydroprocessing catalyst constricts during operation is lacking. Planar anodic alumina films on the other hand with their ideal pore structure (straight, nonintersecting pores) are well suited for analysis by various surface techniques (SEM, EDAX, TEM and XPS) and allow one to visualize the aging process at the single pore level. Experiments of this kind utilizing the anodic films as probes for monitoring catalyst deactivation have been performed by our group (24,46). In one series of experiments anodic alumina films were placed at various positions inside a hydrotreater (at the Unocal Science and Technology Division of Unocal Corporation) containing a Unocal proprietary catalyst, used to upgrade a heavy Arabian atmospheric residuum (11.8 API gravity, 31 ppm nickel, 102 ppm vanadium, 3.9 wt % sulfur, 0.3 wt % nitrogen and 13 wt % asphatenes) under typical hydrotreating conditions. The run lasted for 60 days, after which the catalyst bed was hydrogen stripped and then dumped in sections to locate the anodic alumina samples, which were subsequently analyzed by SEM and EDAX. Fig. 3 shows the SEM and EDAX analysis results of one of these films after the end of the run. Several other films were also collected and analyzed and the results have been presented in one of our recent publications (24). The main conclusions of this study were:

(i) Anodic alumina membranes were shown to be suitable probes for investigating catalyst deactivation during heavy oil hydroprocessing. They provide significant insight into the aging phenomena occuring in real catalysts.

(ii) In particular it was shown that the catalyst experiences the heaviest coking at the reactor inlet. Pores of anodic films with diameters as large as 1000 Å (typical of the macropores of industrial catalysts) were shown to block completely. A maximum in the nickel and vanadium content occured at a point downstream of the reactor. This observation is consistent with a sequential hydrodemetalation mechanism previously proposed for the reaction over real commercial catalysts (52). Following the high metal deposition region and at the bed outlet the catalyst surface is clean with little metal or carbonaceous deposites as measured by SEM/EDAX.

In another series of experiments the anodic aluminas were placed in a 1 liter stirred autoclave (the reaction system is described in (53)) in which the model compound nickel etio-porphyrin (Ni-EP) was hydroprocessed. This model compound is thought to be a good representative of the porphyrinic species present in the heavy crudes and its transport properties have been studied by the use of unimodal membranes by our group (47). The goal of these experiments was to again utilize the anodic aluminas as probes for studies of catalyst deactivation during metaloporphyrin hydrodemetalation and to try to correlate these results with results previously obtained with these aluminas during hydroprocessing of heavy oils. The detailed results of these studies are presented elsewhere (46). For the purposes of the discussion here it suffices to say that the use of anodic alumina membranes with straight nonintersecting pores has enabled us to get a unique insight into the mechanism of deactivation due to metal deposition at the single pore level. Fig. 4 for example shows a TEM photograph of one of those anodic films aged in the autoclave, which hydroprocesses the porphyrin. Note the presence of metal crystallites inside the pore. To the best of our knowledge this is the first time anyone has been able to observe metal crystallites as they deposit during HDM reaction inside a single catalyst pore.

Studies of Transport of Liquid Macromolecules

Anodic alumina membranes are also very valuable in the measurement of transport properties of oil and coal liquid macromolecules. These molecules have sizes, which are comparable with the average pore size of typical hydroprocessing catalysts and as a result transport in these pores occurs by a configurational (hindered) diffusion mechanism The current understanding of how large liquid macromolecules diffuse through restricted porous environments is still rather incomplete and the subject of past and ongoing experimental and theoretical investigations by our group (47-50). Unimodal porous membranes with straight nonintersecting pores (such as the anodic membranes) are ideally suited for such investigations. On the other hand commercial supports, which have tortuous porous structures with distributed pore sizes are not well suited for fundamental investigations of transport and reactions of liquid macromolecules. In typical diffusion experiments with commercial catalysts one finds, for example, that the tortuosity is a strong function not only of the catalyst's properties but also of the size of the diffusing molecule itself. This, of course, is an artifact introduced by the hindered nature of transport and the tortuous catalyst pore structure, one of the several artifacts which make commercial catalysts supports unsuitable for fundamental reaction and transport investigations. The problem is further complicated by the fact that oil and coal liquids are polydisperse molecular systems (a typical petroleum or coal derived asphaltene, as an example, is characterized by a molecule weight range several orders of magnitude broad (48)). Transport of polydisperse macromolecular systems in restricted pores is an interesting phenomenon, which has been studied by our group with the help of unimodal porous membranes (48-50). Under appropriate conditions one observes phenomena similar to chromatographic separation. Exciting possibilities exist in the use of ceramic membrane reactors for hydroprocessing of petroleum and coal derived liquids and work in this area is currently in progress in our laboratory.

Membrane Reactor Studies

As it was already mentioned in the introduction ceramic membranes offer some unique possibilities for high temperature catalytic reaction applications especially for equilibrium limited reactions. In our laboratories we are investigating a number of vapor and gas-phase dehydrogenation reactions. Fig. 5 shows a schematic of our membrane reactor system for studying vapor-phase dehydrogenation reactions. The membrane reactor is shown in Fig. 6. The reactor utilizes flat disk membranes (anodic or Sol-Gel composite-type membranes). To avoid cracking of the membrane due to different thermal expansion coefficients between the membrane and and the reactor walls the reactor system is equipped with a pneumatic arrangement, which compresses the membrane after the appropriate operating temperature has been reached and the temperature of all reactor parts has reached a steady state value.

Fig. 7 shows a schematic of our membrane reactor system for gas-phase reactor applications. It utilizes tubular composite Sol-Gel type membranes commercially available by ALCOA and Sol-Gel composite-type membranes prepared by our group. For sealing, it utilizes compressible Knudsen flow-type seals. The reactor system can operate at temperatures up to 600°C. A number of catalytic hydrocarbon dehydrogenation reactions have been tested in this reactor. In some instances conversions over an order of magnitude higher than the corresponding equilibrium conversions have been observed (43). These are interesting observations from an academic viewpoint but it is unclear to us at this point what the implications are for industrial applications. All catalytic dehydrogenation reactions we have investigated (the cyclohexane dehydrogenation over Pt/ Al_2O_3 catalysts being a typical example (42)) are endothermic and therefore equilibrium conversions increase with temperature as also is the case with the reaction rate. The region of temperatures for the equilibrium conversion to range from approximately 0 to 100% for most of these reactions is fairly narrow and the potential gains therefore in terms of lower operating temperatures are rather small. For such cases the use of catalytic membrane reactors could conceivably be of value only if the catalyst bed exhibited rapid catalyst deactivation at higher temperatures.

In conlusion we have presented here a brief overview of our ongoing studies of transport and reaction phenomena through the use of ceramic membranes. Ceramic membranes have already proven very useful in fundamental investigations of diffusion and reaction of liquid macromolecules and as probes for studying catalyst deactivation at the single pore level. High temperature ceramic membranes also show promise for applications in catalysis and reaction engineering. Preliminary experiments have been encouraging but catalytic membrane reactor technology has still not reached a satisfactory level of maturity. Much remains to be learned before ceramic membrane reactors will find widespread industrial use and acceptance, especially issues of long term hydrothermal stability, resistance to coking and deactivation, mechanical strength and compatability with with other reactor materials.

LITERATURE CITED

1. Hench, L. L. and D. R. Ulrich, Eds., "Science of Ceramic Chemical Processing", Wiley, 1986.

2. Hsieh, H. P., AIChE Symp. Ser., 261, 1, (1988).

3. "Ceramics Make Strong Gid for Tough Membrane Uses", Chemical Engineering, June 9, pp 19, (1986).

4. Hsieh, H. P., R. R. Bhave and H. L. Fleming, J. Membrane Sci., 39, 221, (1988).

5. Kameyama, T., M. Dokiya, K. Fukuda and K. Kotera, Sep. Sci. Technol., 14, 953, (1979).

6. Kameyama, T., M. Dokiya, M. Fujishige, H. Yokokawa and K. Fukuda, Ind. Eng. Chem. Fundam., 20, 97, (1981).

7. Shinji, O., Bull. Chem. Soc. Japan, 55, 2760, (1982).

8. Sun, Y. M. and S. J. Khang, Ind. Eng.

Chem. Res., 27, 1136, (1988).

9. Gryaznov, V. M., Dokl. Akad. Nauk (SSSR) 189, 794, (1969).

10. Gryaznov, V. M., et al., U. S. Patent 4,388,479, (1983); British Patent 1,342,869, (1974); U.S. Patent 3,849,076, (1974); British Patent 2,056,043A, (1981); British Patent 1,528,710, (1978).

11. Gryaznov, V. M., V. S. Smirnov and M. G. Slinko, Proc. 7th Int. Congr. on Cat., (Tokyo), 2, 224, (1981).

12. Gryaznov, V. M., M. M. Ermilova, L. S. Morozova, W. V. Orekhova, V. P. Polyakova, N. R. Roshan, E. M. Savitsky and N. I. Parfenova, J. Less Comm. Met., 89, 529, (1983).

13. Gryaznov, V. M., Platinum Metals Rev., 30, 68, (1986).

14. Gryaznov, V. M., Vestn. Akad. Nauk (SSSR) 3, 21, (1986).

15. Di Cosimo, R., J. D. Burrington and R. K. Grasseli, J. Catal., 102, 234, (1986), also U. S. Patent, 4,571,443, (1986).

16. Michaels, J. N. and C. G. Vayenas, J. Catal., 85, 477, (1984).

17. Nigara, Y. and B. Gales, Bull. Chem. Soc. (Japan) 59, 1997, (1986).

18. Croopnick, G. A. and D. M. Scruggs, U. S. Patent 4,608,319, (1986).

19. Leenaars, A. F. M., K. Keizer and A. J. Burggraaf, J. Mater. Sci., 19, 1977, (1984).

20. Leenaars, A. F. M. and A. J. Burggraaf, J. Membr. Sci., 24, 261, (1985).

21. Leenaars, A. F. M. and A. J. Burggraaf, J. Membr. Sci., 24, 245, (1985).

22. Leenaars, A. F. M. and A. J. Burggraaf, J. Coll. Int. Sci., 105, 27, (1985).

23. Nourbaksh, N., I. A. Webster and T. T. Tsotsis, ACS Symp. Ser., American Chemical Society, 33(3), 502, (Los Angeles) (1988).

24. Nourbaksh, N., I. A. Webster and T. T. Tsotsis, Appl. Catal., 50(1), 65, (1989).

25. Visca, M. and E. Matijenic, J. Colloid Interf. Sci., 68, 308, (1979).

26. Itaya, K., S. Sugawara, K. Arai and S. Saito, J. Chem. Eng. (Japan) 17, 514, (1984).

27. Mitrovic, M. and L. Knezic, Desalination, 28, 147, (1979).

28. Quinn, J. A., J. L. Anderson, W. S. Ho and W. J. Petzny, Biophys. J., 12, 990, (1972).

29. Riedel, C., and R. Spohr, J. Membr. Sci., 7, 225, (1980).

30. Lee, K. H. and S. J. Khang, Ceram. Eng. Sci. Proc., 8, 85, (1987).

31. Koresh, J. E. and A. Sofer, Separ. Sci. Technol., 18, 273, (1983).

32. Wachtman, J. B. and R. A. Haber, Chem. Eng. Prog., 39, (1986).

33. Asadea, M., L. D. Du and M. Ushijima, Proc. 4th Int. Drying Symp., 2, 472, (1984).

34. Asadea, M., L. D. Du, J. Chem. Eng. (Japan) 19, 72, (1986).

35. Asadea, M., L. D. Du and M. Fuji, J. Chem. Eng. (Japan) 19, 84, (1986).

36. Itaya, K., S. Sugawara, K. Arai and Saito, J. Chem. Eng. (Japan) 17, 514, (1984).

37. Saito, M., Bull. Chem. Soc. (Japan) 55, 2760, (1982).

38. Galaj, S., A. Wicker, J. P. Dumas, J. Gillot and D. Garcera, Le Lait, 64, 129, (1984).

39. Poirier, D., F. Maris, M. Bennasar, J. Gillot, D. Garcera and B. Tarodo, Ind. Aliment. Agraic., 101, 1, (1984).

40. Guizard, C., J. A. Alary, A. Larbot, L. Cot, M. Rumeaau, B. Castelas and J. Gillot, Le Lait, 64, 276, (1984).

41. Bhave, R. R. and H. L. Fleming, AIChE Symp. Ser., 261, 19, (1988).

42. Nourbaksh, I. A. Webster and T. T. Tsotsis, paper presented at the AIChE

Annual Meeting, Washington, D.C., Nov. (1988).

43. Champagnie, A. M., I. A. Webster and T. T. Tsotsis, paper presented at the ACS Meeting, Dallas, TX, April 1989; also at the 11th North American Meeting of the Catalysis Society, Detroit, May (1988); also submitted for publication.

44. Abe, F., Eur. Pat. Appl., 228, 885 (1987).

45. Zaspalis, V. T., K. Keizer, J. G. Van Ommen, J. R. H. Ross and A. J. Burggraaf, paper, presented at the AIChE National Meeting, Denver, Aug. (1988).

46. Nourbakhsh, N., I. A. Webster, B. J. Smith, J. Wei and T. T. Tsotsis, paper submitted for publication.

47. Tsotsis, T. T., R. C. Sane, I. A. Webster and J. D. Goddard, J. Catal., 101, 416 (1986)

48. Sane, R. C., I. A. Webster and T. T. Tsotsis, Stud. Surf. Sci. Cat., 36, 165 (1988).

49. Sane, R. C., I. A. Webster and T. T. Tsotsis, ACS Symposium Ser., American Chemical Society, 93(3), 237 (1988).

50. Sane, R. C., I. A. Webster and T. T. Tsotsis, AIChE Symp. Ser., 84, 57 (1988).

51. O'Sullivan, J. P. and G. C. Wood, Proc. Roy. Soc. London, A317, 511 (1970).

52. Ware, R. A. and J. Wei, J. Catal., 93, 100, (1985); also ibid., 93, 121 (1985).

53. Hung, C. W. and J. Wei, Ind. Eng. Chem. Process. Des. Dev., 19, 250 (1980).

ACKNOWLEDGMENT:

We wish to acknowledge the Donors of the Petroleum Research Fund, the U. S. Department of Energy, the USC Faculty Research Fund and the Unocal Corporation for support of this work.

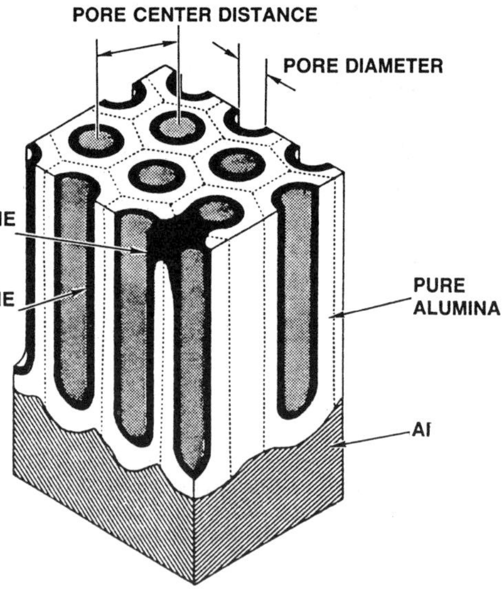

Figure 1. Idealized structure of a porous anodic alumina film.

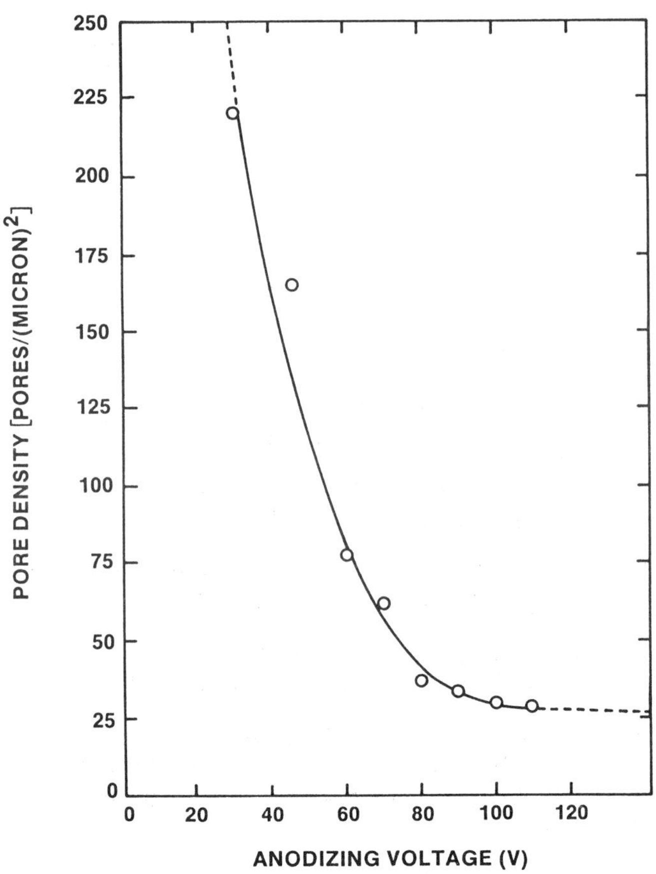

Figure 2. Pore density versus anodizing voltage.

RUN 3IW139, HEAVY ARABIAN AR,
SAMPLE POSITION: R1, MIDDLE (SAMPLE 282-11)

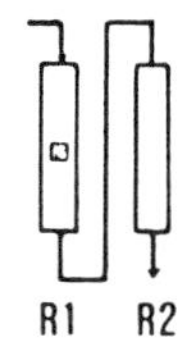

(A)

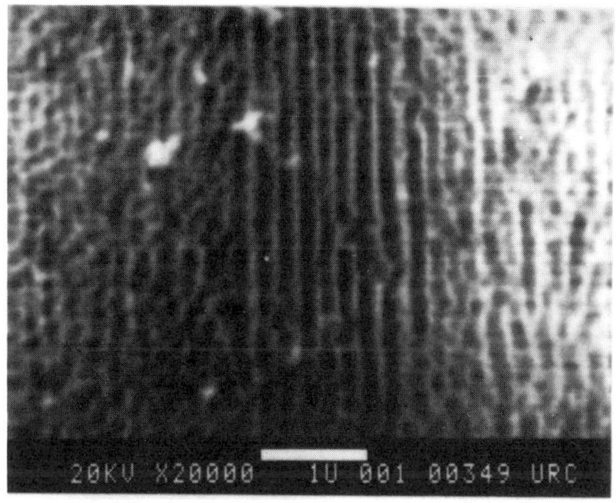

(B)

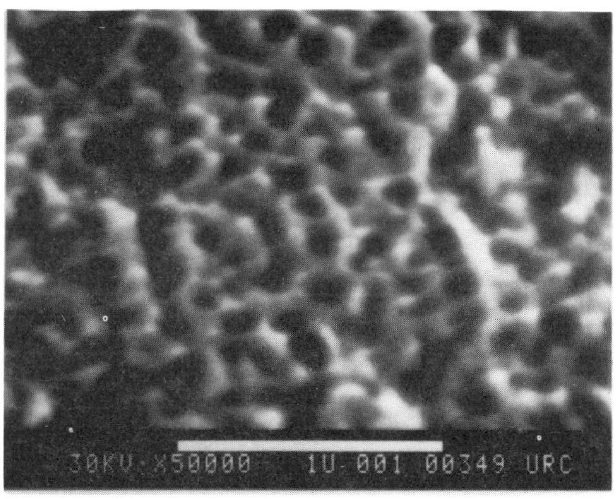

(C)

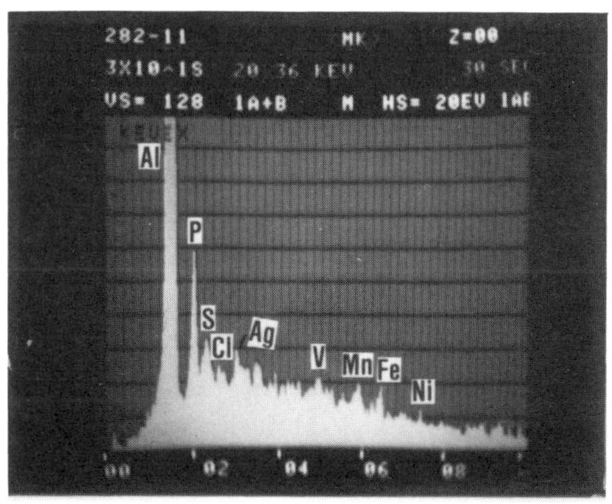

(D)

Figure 3. SEM micrographs and EDAX analysis of an aged porous anodic alumina film.

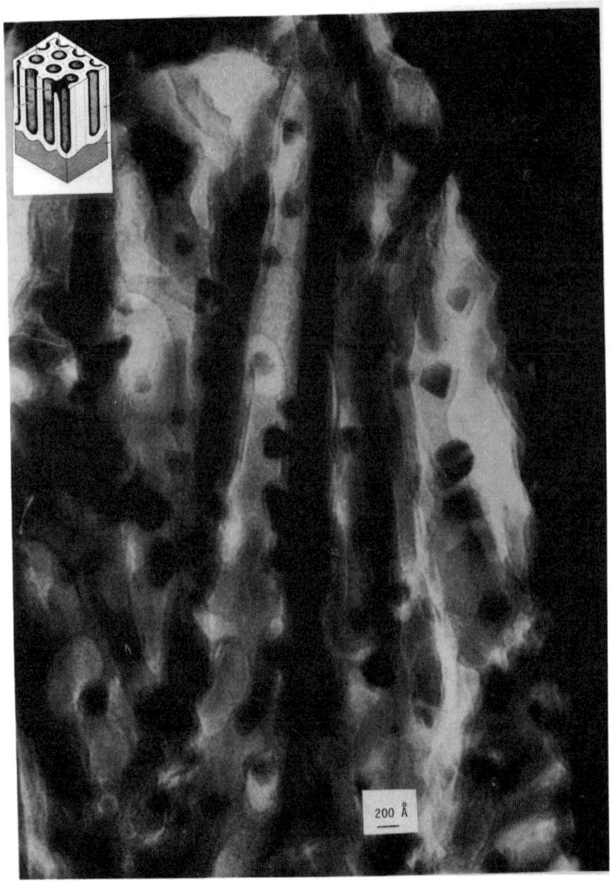

Figure 4. TEM micrograph of an aged anodic alumina film.

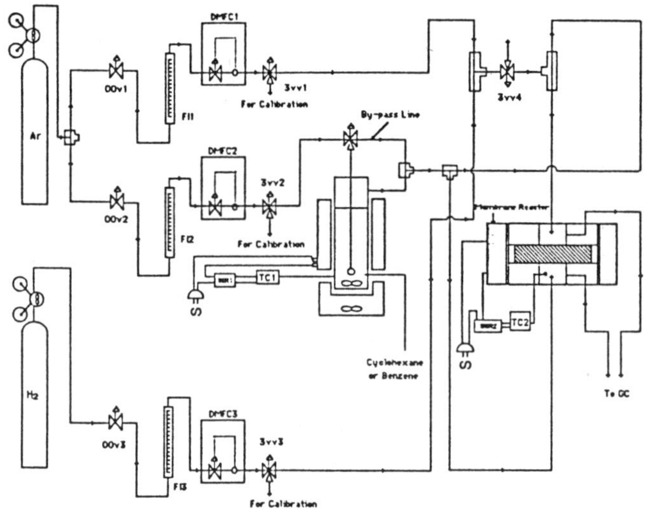

3vv = Three way valve
OOv = On and Off valve
FI = Flow Indicator
DMFC = Digital Mass Flow Controller
TC = Temperature Controller
SSR = Solid State Relay

Figure 5. Membrane reactor system for vapor-phase reaction.

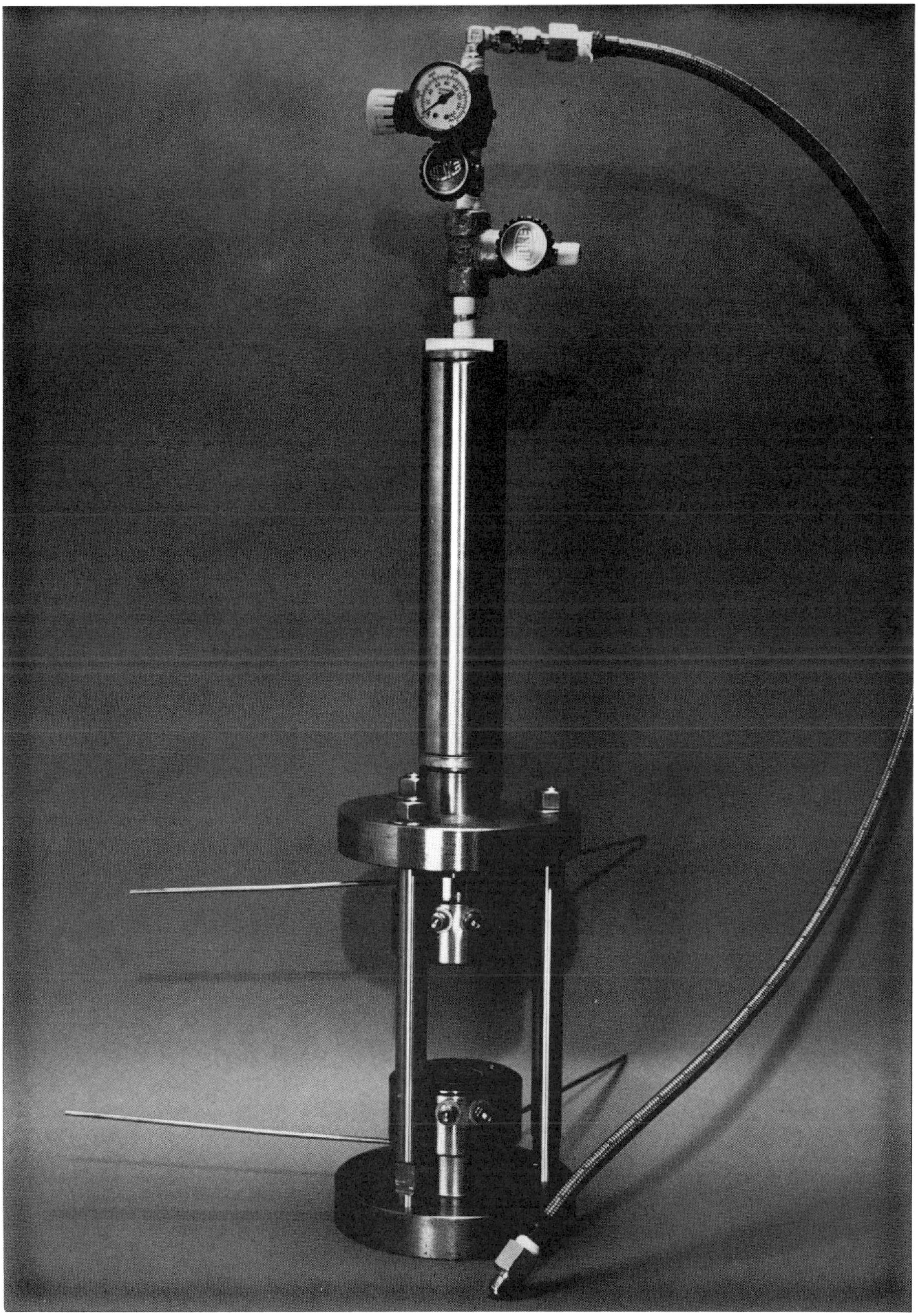

Figure 6. Membrane reactor using flat membranes.

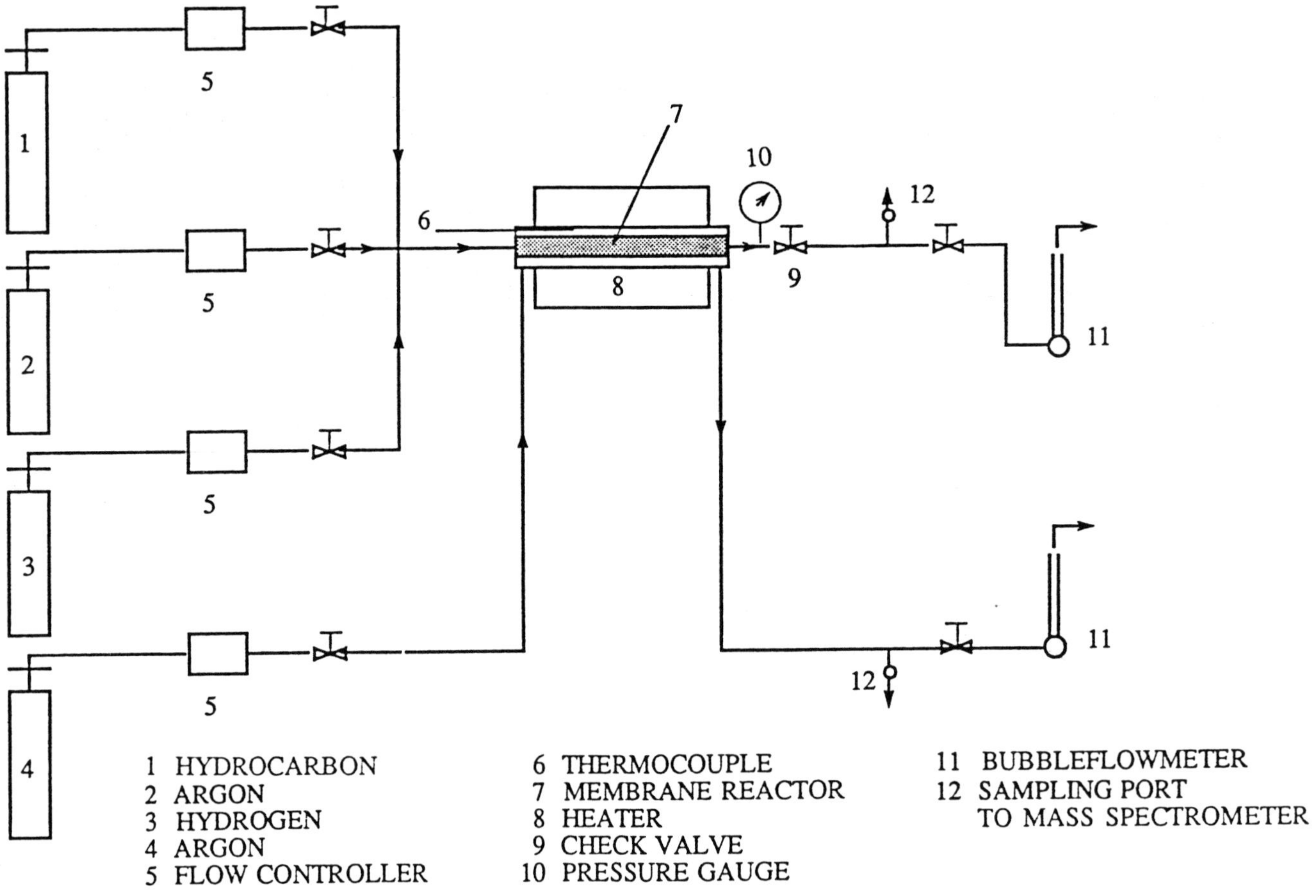

Figure 7. Membrane reactor system for gas-phase applications.

PECTIN DEGRADATION IN A SPIRAL MEMBRANE REACTOR

Xinming Shao, Ying Feng, Shangxu Hu and Rakesh Govind* ■ Dept. of Chemical Engineering, University of Cincinnati, Cincinnati, OH 45221

Pectin degradation by immobilized pectinase has been investigated by pumping sodium polypectate as substrate continuously through a spiral reactor filled by a roll of membrane sheet. The reaction has been conducted at pH = 4.5 and temperature at 40 °C. Substrates containing 0.25, 0.5 and 0.75 mg/ml sodium polypectate have been used in the experiments. Pectin conversion as high as 80% has been achieved at proper operating conditions. Based on the study of free pectinase kinetics, a hypothetical general reaction mechanistic model and the corresponding Michaelis-Menten type rate expression have been proposed for immobilized pectinase. A plug flow reactor model has been suggested for modeling the spiral reactor on the basis of residence time distribution analysis. By best fitting the conversion data to the solutions of the proposed plug flow model combining the Michaelis-Menten type rate expression, a reaction mechanism which includes both substrate and competitive product inhibitions has been identified. Constants involved in the identified mechanism have also been evaluated and used to model the performance of pectin degradation in the spiral reactor.

Most fruits contain a number and variety of pectins in varying concentrations. Elimination of the pectin is the main objective of juice clarification which is required for downstream processes like soluble extraction, pressing, filtration and settling to prepare concentrated products. It has been shown that the most cost-effective way is to degrade the pectin through the use of pectinolytic enzymes (1,2,3). Enzymatic degradation of the pectin allows aggregation and settling out of colloidal material which in the fresh juice is held in suspension. Since 1930's, preparations containing pectinesterase and polygalacturonase enzymes from fungal sources have been used for the expression and the clarification of apple juice (4). The amount of the enzymes used is not high, but the fungal protein and methanol left in the finished product conceivably represent unnecessary health hazards (5). Considerable amounts of commercial enzymes from different sources are employed in the clarification of fruit juices, and almost all of them are some combination of many pectinolytic activities with proper amounts of polymethylgalacturonase (PMG), polygalacturonase (endo and exo PG), pectin methyl esterase (PE) and pectin transeliminase (PTE). In contrast to the rapid expansion of enzyme sources, detailed reaction kinetics and suitable reactors are not well developed and are far from adequate for application.

Retaining enzymes in a reactor system as long as possible is important to reduce the unnecessary health hazards as well as the process cost. Numerous immobilization techniques are developed to attach enzymes to a solid support, for which, porous glass, ion-exchange resins, polyacrylamide and collagen are often used. The activity of enzymes is generally affected by immobilization. At normal operating conditions, immobilized enzymes tend to be less active than their free counterparts. The configurational restrictions and mass transfer limitations while binding on the support might be the main reason for deactivation. The temperature and pH optimums are also observed to shift upon immobilization. Sheiman (5) has immobilized pectin lyase from aspergillus japonicus on collagen films and incorporated the films into a plug flow reactor. The shift of optimum activity from pH 6.0 (free counterparts) to 4.5 (enzyme-collagen film) has been observed (5).

Reactor configuration and operating conditions are other important aspects for commercial application of an immobilized enzyme system. The plug flow reactor which Sheiman (5) used was found to be capable of continuous clarification of fresh apple juice; however the enzymatic activity eventually ceased possibly because of the clogging of the membrane by precipitated pectic materials or by tannin inhibition. Due to the large surface area involved for immobilization and the minimized possibility of plugging of flow channels, spiral reactor is a prominent choice among several available reactor configurations for the enzymatic reaction. It becomes very important especially when the feed is not free of particulate materials. Peterson (6) has recently reported lactose hydrolysis by immobilized β-galactosidase in a capillary bed reactor and a spiral reactor. The latter was found to be better for practical purpose due to the ability of processing particulate materials (6).

* Correspondence should be addressed to Dr. R. Govind

This work involves the investigation of enzyme kinetics of a pectinase complex derived from aspergillus niger free and immobilized forms. Spiral reactor with immobilized enzyme is used for the degradation of pectin suspension. The aim of the research is to identify the immobilized pectinase degradation mechanism and to develop a mathematical expression which can satisfactorily delineate kinetic behavior of the spiral reactor under the given temperature and pH. A hypothetical general mechanistic model is proposed based on the kinetic study of free pectinase. System identification procedure is used for the development and verification of the rate expression as well as the reactor model by best fitting experimental data. Research efforts are devoted to characterize the enzymatic degradation of the pectin in the spiral reactor and to the evaluation of an appropriate reactor model as well as operating conditions.

MATERIALS AND METHODS

Materials

The enzyme used in this study is BIOPECTINASE 700P provided by BICON (U.S.) Inc., which is one of the commercially available powder pectinases with high activity and ideal for the elimination of pectin in fruit, wine or other industries. It's working pH ranges from 2.0 to 6.2 with an optimum between 3.0 to 5.5, and it can in general perform between 2 to 65°C with an optimum around 54 to 60°C (7). Referring to the preferable reaction conditions recommended, experiments in this study were carried out at T= 42°C and pH = 4.6.

In order to obtain consistent experimental results, sodium salt of polygalacturonic acid, or sodium polypectate (P-1879, SIGMA), rather than natural fruit juice was used as substrate in this work. The product of this enzyme reaction is d-galacturonic acid. All chemicals were obtained from commercial sources and all were reagent grade or better. Deionized water from Fisher Scientific Co. was used exclusively in all experiments.

The spiral reactor module containing immobilized pectinase which was identical to BIOPECTINASE 700P was manufactured by Amerace Corporation (Hackettstown, NJ). The enzyme was immobilized on silica particles which were physically entrapped in a ribbed sheet of microporous polyvinylchloride. The sheet was rolled in a jelly-roll configuration which was epoxied to two plastic plates and housed in a casing of plexiglas and stainless steel. The feedstock which was sodium polypectate solution in our case was pumped through the spiral direction channels formed by the ribs. A cross view of the spiral reactor configuration is shown in Figure 1.

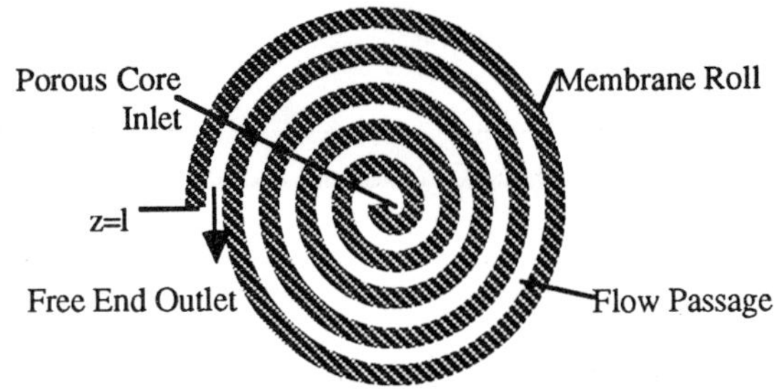

Figure 1: A cross section view of the spiral reactor configuration

Methods and Procedures

Two types of enzyme kinetic studies were performed in this work. Initially, studies of free counterparts of pectinase was conducted in a batch reactor with different substrate concentrations, different amounts of pectinase and different reaction times to explore the effects of each of these factors.

In addition, studies of immobilized pectinase were carried out using a spiral reactor. A schematic drawing of the apparatus utilized in the kinetics studies of immobilized enzyme in the spiral reactor is shown in Figure 2. The spiral reactor module was totally submerged in a constant temperature water bath which was controlled at 42°C for all experiments.

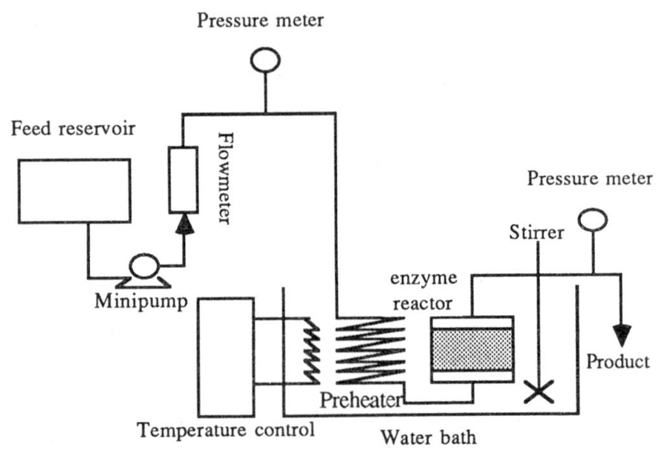

Figure 2: Schematic drawing of the apparatus utilized in the kinetic studies of immobilized enzyme in the spiral reactor

At a given flow rate, data on pectin conversion was obtained via determination of the concentration of d-galacturonic acid present in the effluent stream with an established steady state. A spectrophotometer

(SPECTRONIC 1001, MILTON ROY Co.) with the suggested DNS analysis procedure (7) was used for determination of the concentration of d-galacturonic acid. The effluent concentration of d-galacturonic acid was also monitored to check whether a steady-state was achieved. Normally, 4 to 6 reactor space times were required to reach a steady state. After the samples were collected and analyzed, the flowrate was changed to achieve a new steady state. This procedure was repeated until the desired number of flowrates were taken.

Data from the spiral reactor were collected for several experiments with different flowrates during an whole day operation. Then the reactor was cleaned by flushing sequentially with 1.5 liter of 0.05M sodium acetate buffer (pH=4.5). The reactor module was then filled with 200ppm 0.05M acetate buffer (pH=4.5) with 0.02% sodium azide in it to prevent microbial growth and stored overnight at 4°C. After that, the experimental system with all pumps and tubes was cleaned by pumping at least 1 liter of deionized water.

Residence time distribution (RTD) study was performed for the investigation of the nature of mixing in the spiral module. The primary spiral direction flow pattern coupling with the semi-porous diffusion through the supported material complicated the assessing of the magnitude of the dispersion parameter in the reactor. In order to conduct the RTD study, several modifications were made to the experimental system shown in Figure 2. The long tubing for pre-heating and the water bath were removed and the reactor module was installed directly after the pump. Pulses of concentration of sodium acetate buffer were employed and a conductivity electrode connected to an AMI model impedance meter was used to record the pulses of concentration in inlet and the response in outlet.

RESULTS AND DISSCUSION

Kinetic Studies on Free Pectinase

The basic experiments are conducted to obtain a number of time-yield data, which shows the variation of enzyme reaction product concentration C_P with the concentration of enzyme C_E, the concentration of substrate C_S, and the reaction time t. Results of a set of such experiments taken at C_E = 0.2 (mg/ml) is shown in Figure 3. Other sets of experiments were taken at C_E = 0.1, 0.3, 0.5, 0.6, 0.7, 0.8, 0.9, 1.0 etc.

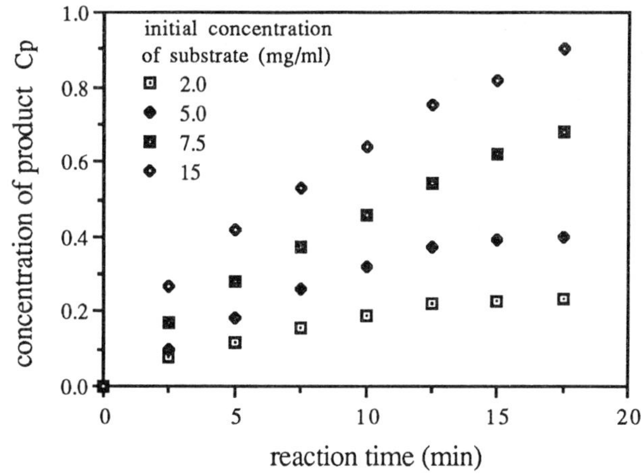

Figure 3: Time-yield curves at C_E = 0.2 (mg/l)

Plots of reciprocal initial reaction rate $1/r_0$ verses reciprocal substrate concentration $1/C_S$ were made. A plot for C_E = 0.2 (mg/l) is shown in Figure 4.

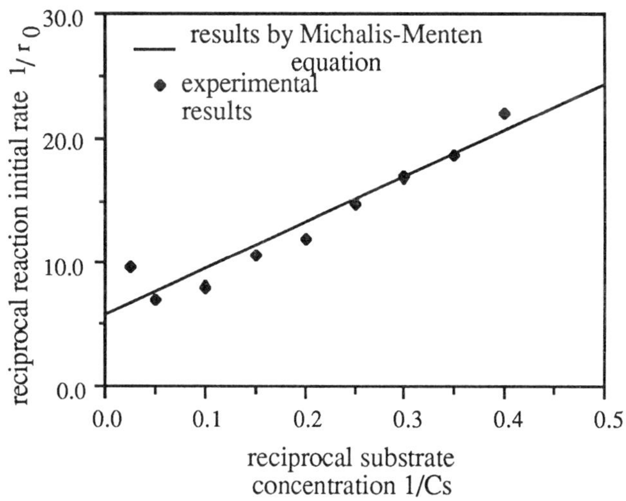

Figure 4: Plot of reciprocal initial reaction rate $1/r_0$ verses reciprocal substrate concentration $1/C_S$ at C_E = 0.2 (mg/l)

The solid straight line in Figure 4 represents the prediction of Michaelis-Menten equation:

$$\frac{1}{r_0} = \frac{1}{r_{0m}} + \frac{K_m}{r_{0m} C_S} \tag{1}$$

or

$$r_0 = \frac{r_{0m} C_S}{K_m + C_S} \tag{2}$$

It can be found there exists a local minimum and an asymptotic approach to the y-axis on our Lineweaver-Burke plot (Figure 4). In other words, our reaction system exhibits a substrate inhibition mechanism.(8).

However, the simulation results obtained by only including the substrate inhibition for the concentration of product as a function of reaction time gives a curve which has quite a different shape than that of the experimental time-yield data shown in Figure 3. The discrepancy between the simulation results and the experimental data hints certain type of inhibition shown by the product toward the reaction (9). This effect might be included in the kinetic equation.

Reaction Mechanism and Kinetic Equation for Immobilized Pectinase

On the basis of the results obtained in the kinetic studies on free pectinase, a hypothetical general reaction mechanistic model of the Michaelis-Menten type has been suggested for pectin degradation. The model accounts for the substrate inhibition, the competitive and the noncompetitive product inhibitions. The hypothetical general reaction mechanism can be schematically illustrated as follows:

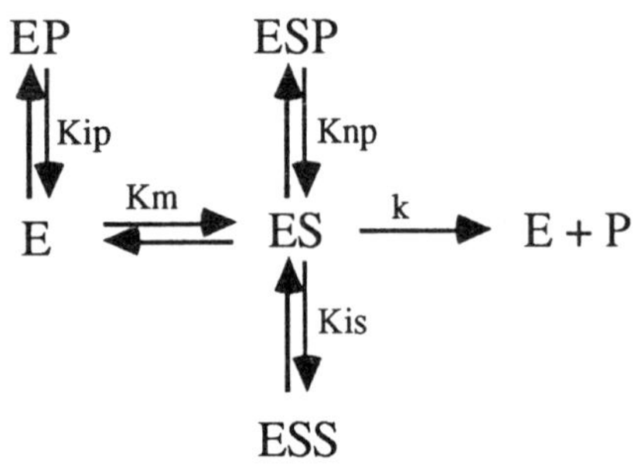

(3)

where E, S, P are the free enzyme, substrate (sodium polypectate), product (d-galacturonic acid) respectively, and ES, EP, ESS ESP are the enzyme-substrate complex, enzyme-product complexes. Among these complexes, only ES is active. K_{ip}, K_{np}, K_m, K_{is} are the dissociation equilibrium constants for each process and k is the rate constant.

Defining the following dimensionless groups,

$$X_S = C_S / S_0,\ X_P = C_P / S_0 \qquad (4)$$

the corresponding rate expression can be derived with the usual steady state assumption for the concentration of the enzyme-substrate complex (ES) and the conservation of total enzyme.

$$r_s = -\frac{V_m X_S S_0}{\frac{(X_S S_0)^2}{K_{is}} + (1+\frac{X_p S_0}{K_{np}}) X_S S_0 + (1+\frac{X_p S_0}{K_{ip}}) K_m} \qquad (5)$$

When there is no inhibition, K_{ip}, K_{np}, K_{is} approach infinity, then the rate expression reduces to the Michaelis-Menten equation. This special case with the other three cases are summarized in Table 1.

Table 1: Summary of special rate expressions derived from the hypothetical general mechanistic model (equation 5)

no inhibition	
$r_s = -\frac{V_m X_S S_0}{X_S S_0 + K_m}$	(6)
substrate inhibition	
$r_s = -\frac{V_m X_S S_0}{\frac{(X_S S_0)^2}{K_{is}} + X_S S_0 + K_m}$	(7)
substrate and competitive product inhibition	
$r_s = -\frac{V_m X_S S_0}{\frac{(X_S S_0)^2}{K_{is}} + X_S S_0 + (1+\frac{X_p S_0}{K_{ip}}) K_m}$	(8)
substrate and noncompetitive product inhibition	
$r_s = -\frac{V_m X_S S_0}{\frac{(X_S S_0)^2}{K_{is}} + (1+\frac{X_p S_0}{K_{np}}) X_S S_0 + K_m}$	(9)

Analysis of residence time distribution

For an open system with small backmixing, the dispersion coefficient (D) may be related to changes in variances (σ^2) of a pulse entering and exiting the reactor and the mean residence time (t_m) in the following form (10):

$$8\left(\frac{D}{uL}\right)^2 + 2\frac{D}{uL} = \frac{\sigma^2_{out} - \sigma^2_{in}}{t_m} \qquad (10)$$

Assuming that the volumetric flow rate is a constant the values of σ^2 and t_m can be determined from experimental data with the following formulas:

$$t_m = \frac{\Sigma t_i c_i \Delta t}{\Sigma c_i \Delta t} \qquad (11)$$

and

$$\sigma^2 = \frac{\Sigma t_i^2 c_i \Delta t}{\Sigma c_i \Delta t} - t_m^2 \qquad (12)$$

Table 2 lists values of the term (D/uL) calculated by the above simple approach based on equation (10) to (12) at various reactor space time. There are some other methods available for the estimation of the term (11). However, the term normally is used to determine the magnitude of departure of a real reactor from a plug flow model, it is not necessary to know this term to a high degree of accuracy. The inverse form of this term is known as the Peclet number which implies the tendency of a real reactor approaching plug flow conditions.

Table 2: Results of residence time distribution analysis

Reactor Space Time (min.)	D/uL
25.3	0.0212
20.2	0.0186
14.6	0.0232
8.8	0.0213
5.5	0.0251
2.8	0.0234
0.9	0.0252

The results listed in Table 2 shows that the magnitude of the Peclet number for reactor space time of interest (1 to 25 min.) ranges around 10^2, which indicates that the departure of the spiral reactor from plug flow behavior is not large (i.e., D/uL around 0.025). Figure 5 is one plot of the experimental result of the RTD study.

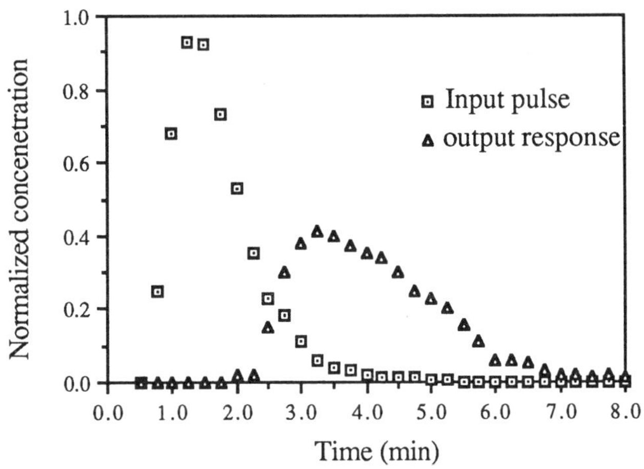

Figure 5: Result of RTD experiment where:
normalized concentration:

$$C_N = \frac{|\,C_s - S_{01}\,|}{|\,S_{01} - S_{02}\,|}$$

space time τ = 12.5 min,
initial substrate concentration S_{01} = 7.5
S_{02} = 2.5 mg/ml

Modeling the spiral reactor

The result obtained in the RTD study implies that we may simplify the mathematical model for the spiral reactor to a plug flow reactor which is in the form of first order differential equation:

$$\frac{dX_s}{d\tau} = \frac{r_s}{S_0} \tag{13}$$

In the hypothetical general rate expression or in the special cases listed in Table 1 (equations 5 to 9), there are several unknown parameters. In order to estimate these parameters and to access the characteristics of the spiral reactor, three sets of experiments with initial substrate solution concentration = 2.5, 5.0, and 7.5mg/ml were conducted. Pectine conversion data via reactor space time were collected. The experimental results are illustrated in Figure 6. Parameters are identified by fitting the experimental data with the solution of equation 13 and the general rate expression equation 5 in the sense of least squares. The quasilinearization technique with simultaneous calculation of approximation (12,13) is employed in this work. Derivatives evaluation was performed by the automatic derivative evaluation scheme developed by Bellman and collaborators (14).

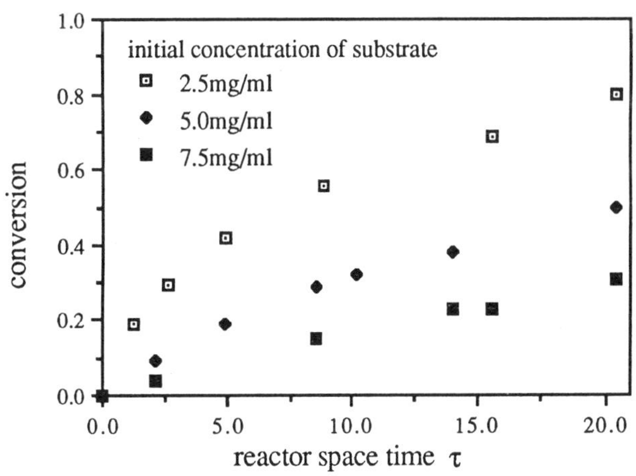

Figure 6: Pectin conversion data verses space reactor time
initial substrate concentration S_0 = 2.5, 5.0, 7.5 mg/ml

Since K_m generally is a sufficiently large term in the denominator of the rate expressions, neglecting those terms without K_m appearing as a factor will be a reasonable approximation. Then the general rate expression becomes:

$$r_s = -\frac{V_m X_s S_0}{(1+\frac{X_p S_0}{K_{ip}})K_m} \tag{14}$$

Identification results based on this rate expression was used for feasible starting points for the parameters V_m/K_m and K_{ip}. This enhanced the effectiveness of the identification algorithms.

In all the three cases with different initial substrate concentrations, the parameter identification indicated that K_{np} was two or three magnitudes larger than other dissociation equilibrium constants, which made the term X_pS_0/K_{np} insignificant. The results suggests a mechanism which only includes the substrate and the competitive product inhibition. The corresponding rate expression is equation 8. Table 3 summarizes the results of parameter identification with equation 8 as the rate expression.

Table 3: Results of parameter identification with the rate expression of equation 8

Parameters	initial concentration of substrate solution 2.5mg/ml	5.0mg/ml	7.5mg/ml
K_m (M)	0.98	0.82	0.93
V_m (M/min)	2.18	1.86	1.72
K_{is} (M, 10^{-5})	0.51	0.43	0.45
K_{ip} (M, 10^{-4})	0.86	0.82	0.76

Figure 7 illustrates the experimental data versus fitting results using equation 5 or equation 8 as rate expressions. It indicates that both equations fit very well, and there is no significant improvement by including the noncompetitive product inhibition as in equation 5 (sums of least squares are around $3.50*10^{-2}$). In the absence of additional information about the enzyme system, the mechanistic model which only includes the substrate and competitive product inhibitions (equation 8) should be the most appropriate. Identification studies were also carried out with equations 6,7 and 9 as rate expressions. None of them were found to compete with the results obtained by using equations 5 or 8.

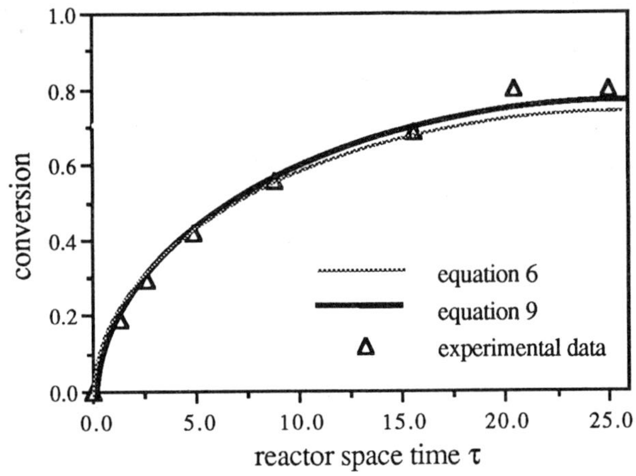

Figure 7: Experimental data versus the fitting results obtained by using equation 6 or equation 9 as rate expressions, initial substrate concentration S_0 = 2.5 mg/ml

It should be noted that the plug flow model is not always valid. It is just a valid simplification in the range of space time we are interested. One-dimensional tubular reactor model with axial dispersion was also considered in this work. However, in most our instances, the difference between the conversions predicted by plug flow model and the axial dispersion model is not significant even though same kinetic parameters are used.

CONCLUSIONS

The spiral reactor studied in this research is quite effective in pectin degradation. Over 80% of the pectin can be eliminated in this reactor in a space time of 15 minutes. The reactor can be cleaned daily without noticeable loss in enzyme activity.

Residence time distribution studies indicated that the Peclet number is around 10^2 within the range of reactor space time of interest. It is shown that the reactor can be approximated by a plug flow reactor without significant error.

A mechanistic kinetic model that includes substrate and competitive product inhibitions is identified by the quasilinearization technique. The identified model can best fit the data with a plug flow reactor design equation. It provides a basis for reactor scale-up which would be appropriate for use in the continuous pectin degradation.

ACKNOWLEDGEMENT

BIOPECTINASE 700P used in this work was supplied by BIOCON Co. and the spiral reactor was provided by Amerace Corporation (Hackettstown, NJ). In addition, the authors would like to express their thanks to Dr. Bruce Goldberg (Amerace Corp.) for providing the reactor on a complimentary basis and Dr. Robert Chen (Amerace Corp.) for technical assistance.

NOTATION

Roman Letters

c_i	tracer concentration at time t_i
C_S, C_P, C_E	concentration for substrate, product, and enzyme
C_N	normalized concentration
E	enzyme (pectinase)
ES	active enzyme-substrate complex
EP	inactive enzyme-product complex
ESS	inactive ternary enzyme-substrate complex
k	reaction rate constants
$K_{ip}, K_{np}, K_m, K_{is}$	dissociation equilibrium constants for each process
L	length of the fluid passage in reactor
P	product (d-galacturonic acid)
r_0, r_{0m}	initial reaction rate and maximum initial reaction rate
r_S	reaction rate of substrate
S	substrate (sodium polypectate)
S_0	initial concentration of substrate
S_{01}, S_{02}	two different initial concentrations of substrate used for RTD study
Δt	time increment
t_i	time of the ith measurement
u	mean velocity of the fluid
X_S, X_P	dimensionless substrate and product concentrations $X_S = C_S / S_0$, $X_P = C_P / S_0$
Z	dimensionless axial location

Greek letters

Δ	increment
τ	reactor space time
σ^2_{in}	variances of the responses to a pulse input measured at the reactor inlet
σ^2_{out}	variances of the responses to a pulse input measured at the reactor outlet

LITERATURE CITED

1. Yamasaki, M., T. Yasui and K. Arima, " Pectic Enzymes in the Clarification of Apple Juice, Part I. Study on the Clarification Reactions in a Simplified Model," Agric. Biol. Chem., 28, 779(1964).

2. Weetall, H. H. Immobilized Enzymes: " Some Applications to Foods and Beverages, Part I, Immobilization Methods," Food Product Develop. 7, 46(1973).

3. Weetall, H. H. " Immobilized Enzymes: Some Applications to Foods and Beverages, Part II, Applications," Food Product Develop. 7, 94(1973).

4. Yamasaki, M., A. Kato and K. Arima, " Pectic Enzymes in the Clarification of Apple Juice, Part II. The Mechanism of Clarification," Agric. Biol. Chem., 31, 552(1967).

5. Sheiman, Mark I.," Properties and Applications of Free and Immobilized Pectic Enzymes," PhD Thesis, Rutgers University, 1974.

6. Peterson, Ralph Scott.," Lactose Hydrolysis by Immobilized β-Galactosidase in a Capillary Bed Reactor," PhD Thesis, University of Wisconsin-Madison, 1987.

7. Product Information, BIOPECTINASE, Biocon pectinase enzyme system, BIOCON (U.S.) Inc., 1988.

8. Bailey James E. and David F. Ollis, Biochemical Engineering Fundamentals, 2nd edition, McGraw-Hill, 1987.

9. Laidler, K. J. and P. S. Bunting, The Chemical Kinetics of Enzyme Action, 2nd edn., Clareudon Press, 1973.

10. Levenspiel, O., Chemical Reaction Engineering, 2nd edn., John Wiley and Sons Inc., New York, 1973.

11. Hill, C. G. Jr., An Introduction to Chemical Reaction Engineering Kinetics and Reactor Design, John Wiley and Sons Inc., New York, 1977.

12. Bellman, R., " Successive Approximations and Computer Storage Problems in Ordinary Differential Equations, " Comm .of the ACM, 4, 222(1965).

13. Bellman, R., H. Kagiwada and R. Kalaba, " Quasilinearization, System Identification, and Prediction, " Int. J. Eng. Sci., 3, 327(1965).

14. Bellman, R., H. Kagiwada and R. Kalaba, " Wengert's Method for Partial Derivatives, Orbit Determination and Quasilinearization, " Comm. of the ACM, 8, 231(1965).

INDEX

SYMPOSIUM SERIES

ADSORPTION

96 Developments in Physical Adsorption
117 Adsorption Technology
219 Recent Advances in Adsorption and Ion Exchange
230 Adsorption and Ion Exchange—'83
233 Adsorption and Ion Exchange—Progress and Future Prospects
242 Adsorption and Ion Exchange: Recent Developments
264 Adsorption and Ion Exchange: Fundaments and Applications

AEROSPACE

33 Rocket and Missile Technology
52 Chemical Engineering Techniques in Aerospace

BIOENGINEERING

69 Bioengineering and Food Processing
84 The Artificial Kidney
86 Bioengineering ... Food
93 Engineering of Unconventional Protein Production
99 Mass Transfer in Biological Systems
108 Food and Bioengineering—Fundamental and Industrial Aspects
114 Advances in Bioengineering
163 Water Removal Processes: Drying and Concentration of Foods and Other Materials
172 Food, pharmaceutical and bioengineering—1976/
181 Biochemical Engineering Renewable Sources of Energy and Chemical Foodstocks

CRYOGENICS

224 Cryogenic Processes and Equipment 1982
251 Cryogenic Properties, Processes and Applications 1986

CRYSTALLIZATION

110 Factors Influencing Size Distribution
193 Design Control and Analysis of Crystallization Processes
215 Nucleation, Growth and Impurity Effects in Crystallization Process Engineering
240 Advances in Crystallization From Solutions
253 Fundametnal Aspects of Crystallization and Precipitation Processes

DRAG REDUCTION

11 Drag Reduction
130 Drag Reduction in Polymer Solutions

ENERGY

Conversion and Transfer

5 Heat Transfer, Atlantic City
57 Heat Transfer, Boston
59 Heat Transfer, Cleveland
75 Energy Conversion Systems
79 Heat Transfer with Phase Change
87 Advances in Cryogenic Heat Transfer
113 Convective and Interfacial Heat Transfer
118 Heat Transfer—Tulsa
119 Commercial Power Generation
138 Heat Transfer—Research and Design
162 Energy and Resource Recover from Industrial and Municipal Solid Wastes
174 Heat Transfer, Research and Application
189 Heat Transfer—San Diego 1979
202 Transport with Chemical Reactions
208 Heat Transfer—Milwaukee 1981
216 Processing of Energy and Metallic Minerals
225 Heat Transfer—Seattle 1983
236 Heat Transfer—Niagra Falls 1984
245 Heat Transfer—Denver 1985
257 Heat Transfer—Pittsburgh 1987
263 Heat Transfer—Houston 1988
269 Heat Transfer—Philadelphia 1989

Nuclear Engineering

53 Part XIII
56 Part XIV
95 Part XX
104 Part XXI
106 Part XXII
119 Commercial Power Generation
168 Heat Transfer in Thermonuclear Power Systems
169 Developments in Uranium Enrichment
191 Nuclear Engineering Questions Power Reprocessing, Waste, Decontamination Fusion
221 Recent Developments in Uranium Enrichment

ENVIRONMENT

78 Water Reuse
97 Water—1969
115 Important Chemical Reactions in Air Pollution Control
122 Chemical Engineering Applications of Solid Waste Treatment
124 Water—1971
126 Air Pollution and its Control
133 Forest Products and the Environment
137 Recent Advances in Air Pollution Control
139 Advances In Processing and Utilization of Forest Products
144 Water—1974: I. Industrial Wastewater Treatment
145 Water—1974: II. Municipal Wastewater Treatment
146 Forest Product Residuals
147 Air: I. Pollution Control and Clean Energy
148 Air: II. Control of NO_{xx} and SO_x Emissions
149 Trace Contaminants in the Environment
151 Water—1975
156 Air Pollution Control and Clean Energy
157 New Horizons for the Chemical Engineer in Pulp and Paper Technology
165 Dispersion and Control of Atmospheric Emissions, New-Energy-Source Pollution Potential
170 Intermaterials Competition in the Management of Shrinking Resources
171 What the Filterman Needs to Know About Filtration
175 Control and Dispersion of Air Pollutants: Emphasis on NO_X and Particulate Emissions
177 Energy and Environmental Concerns in the Forest Products Industry
184 Advances in the Utilization and Processing of Forest Products
188 Control of Emissions from Stationary Combustion Sources Pollutant Detection and Behavior in the Atmosphere
195 The Role of Chemical Engineering in Utilizing the Nation's Forest Resources
196 Implications of the Clean Air Amendments of 1977 and of Energy Considerations for Air Pollution Control
198 Fundamentals and Applications of Solar Energy
200 New Process Alternatives in the Forest Products Industries
201 Emission Control from Stationary Power Sources: Technical, Economic and Environmental Assessments
207 The Use and Processing of Renewable Resources—Chemical Engineering Challenge of the Future
209 Water—1980
210 Fundamentals and Applications of Solar Energy II
211 Research Trends in Air Pollution Control: Scrubbing, Hot Gas Clean-up, Sampling and Analysis
213 Three Mile Island Cleanup
223 Advances in Production of Forest Products
232 Applications of Chemical Engineering in the Forest Products Industry
239 The Impact of Energy and Environmental Concerns on Chemical Engineering in the Forest Products Industry
243 Separation of Heavy Metals and Other Trace Contaminants
246 Advances in Process Analysis and Development in the Forest Products Industries.
265 Resource Recovery of Municipal Solid Wastes

FLUIDIZATION

101 Fundamental Processes in Fluidized Beds
105 Fluidization Fundamentals and Application
116 Fluidization: Fundamental Studies Solid-Fluid Reactions, and Applications
176 Fluidization Application to Coal Conversion Processes
205 Recent Advances in Fluidization and Fluid-Particle Systems
234 Fluidization and Fluid Particle Systems: Theories and Applications
241 Fluidization and Fluid Particle Systems: Recent Advances
255 New Developments in Fluidization and Fluid-Particle Systems
262 Fluidization Engineering: Fundamentals and Applications

HISTORY OF CHEMICAL ENGINEERING

100 The History of Penicillin Production
235 Diamond Jubilee Historical Review Volume

ION EXCHANGE

79 Adsorption and Ion Exchange Separations
219 Recent Advances in Adsorption and Ion Exchange
230 Adsorption and Ion Exchange—'83
233 Adsorption and Ion Exchange—Progress and Future Prospects
259 Recent Progress in Adsorption and Ion Exchange

KINETICS

25 Reaction Kinetics and Unit Operations
73 Kinetics and Catalysis

MINERALS

15 Mineral Engineering Techniques
85 Fossil Hydrocarbon and Mineral Processing
173 Fundamental Aspects of Hydrometallurgical Processes
180 Spinning Wire from Molten Metals
216 Processing of Energy and Metallic Minerals

PETROCHEMICALS

49 Polymer Processing
127 Declining Domestic Reserves—Effect on Petroleum and Petrochemical Industry
135 The Petroleum/Petrochemical Industry and the Ecological Challenge
142 Optimum Use of World Petroleum
212 Interfacial Phenomena in Enhanced Oil Recovery

PETROLEUM PROCESSING

103 C_4 Hydrocarbon Production and Distribution
127 Declining Domestic Reserves—Effect on Petroleum and Petrochemical Industry
135 The Petroleum/Petrochemical Industry and the Ecological Challenge
155 Oil Shale and Tar Sands
226 Underground Coal Gasification: The State of the Art

PHASE EQUILIBRIA

2 Pittsburgh and Houston
6 Collected Research Papers
88 Phase Equilibria and Gas Mixtures Properties

PROCESS DYNAMICS

36 Process Dynamics and Control
46 Process Systems Engineering
55 Process Control and Applied Mathematics
214 Selected Topics on Computer-Aided Process Design and Analysis
267 Process Sensing and Diagnostics

SEPARATION

120 Recent Advances in Separation Techniques
192 Recent Advances in Separation Techniques—II
250 Recent Advances in Separation Techniques—III

SONICS

109 Sonochemical Engineering

MISCELLANEOUS

48 Chemical Engineering Reviews
70 Small-Scale Equipment for Chemical Engineering Laboratories
112 Engineering, Chemistry, and Use of Plasma Reactors
125 Vacuum Technology at Low Temperatures
143 Standardization of Catalyst Test Methods
182 Biorheology
183 The Modern Undergraduate Laboratory Innovative Techniques
185 Electro Organic Synthesis Technology
186 Plasma Chemical Processing
187 Chronic Replacement of Kidney Function
194 Hazardous Chemical—Spills and Waterborne Transportation
203 A Review of AIChE's Design Institute for Physical Property Data (DIPPR) and Worldwide Affiliated Activities
204 Tutorial Lectures in Electrochemical Engineering and Technology
206 Controlled Release Systems
217 New Composite Materials and Technology
220 Uncertainty Analysis for Engineers
228 Problem Solving
229 Tutorial Lectures in Electrochemical Engineering and Technology—II
231 Data Base Implementation and Application
237 Awareness of Information Sources
238 New Developments in Liquid-Liquid Extractors: Selected Papers From ISEC '83
244 Experimental Results from the Design Institute for Physical Property Data. I: Phase Equilibria
247 Chemical Engineering Data Sources
248 Industrial Membrane Processes
249 Measurement of High Temperatures in Furnaces and Processes
252 Thin Liquid Film Phenomena
254 Electrochemical Engineering Applications
256 Experimental Results From the Design Institute for Physical Property Data: Phase Equilibria and Pure Component Properties
258 Fiber Optics: Processing and Applications
261 New Membrane Materials and Processes for Separation
266 Diffusion and Convection in Porous Catalysts
268 Membrane Reactor Technology

MONOGRAPH SERIES

3 The Manufacture of Nitric Acid by the Oxidation of Ammonia—The DuPont Pressure Process by Thomas H. Chilton
4 Experiences and Experiments with Process Dynamics by Joel O. Hougen
5 Present, Past, and Future Property Estimation Techniques by Robert C. Reid
6 Catalysts and Reactors by James Wei
7 The 'Calculated' Loss-of-Coolant Accident by L.J. Ybarrondo, C.W. Solbrig, H.S. Isbin
8 Understanding and Conceiving Chemical Process by C. Judson King
9 Ecosystem Technology: Theory and Practice by Aaron J. Teller
10 Fundamentals of Fire and Explosion by Daniel R. Stull
11 Lumps, Models and Kinetics in Practice by Vern W. Weekman, Jr.
12 Lectures in Atmospheric Chemistry by John H. Seinfeld
13 Advanced Process Engineering by James R. Fair
14 Synfuels from Coal by Bernard S. Lee
15 Computer Modeling of Chemical Processes by J.D. Seader
16 "High-Tech" Materials by Sheldon Isakoff
17 Separations: New Directions for an Old Field
128 Biotechnology: Status and Perspectives